water from heaven

Water from Heaven

THE STORY OF WATER FROM THE BIG BANG TO THE RISE OF CIVILIZATION, AND BEYOND

Robert Kandel

COLUMBIA UNIVERSITY PRESS   NEW YORK

Columbia University Press

Publishers Since 1893

New York   Chichester, West Sussex

Originally published in France as *Les eaux du ciel*

Copyright © 1998 Hachette Littératures;

English-language translation copyright © 2003 Columbia University Press

Library of Congress Cataloging-in-Publication Data

Kandel, Robert S.

[Les eaux du ciel. English]

Water from heaven : the story of water from the big bang to the rise of civilization, and beyond / Robert Kandel.

p.   cm.

Includes bibliographical references and index.

ISBN 0–231–12244–6 (cloth : alk. paper)

1. Water.   I. Title.

GB661.2 .K3513 2003

551.46—dc21

2002031229

Columbia University Press books are printed on permanent and durable acid-free paper.

Printed in the United States of America

c 10 9 8 7 6 5 4 3 2 1

I dedicate this book to Danielle, Maya, Tania, and Arielle,

in the order of their appearance in my life

CONTENTS

PREFACE

Why another book about water? For one thing, water has always been and continues to be one of the major challenges of civilization. It could be the biggest problem of the 21st century. In today's world, farmers, lawyers, decision-makers, scientists, engineers, indeed all citizens must share in this both old and new responsibility. Too often, even when raising the question in terms of "global management," those responsible for this vital resource only consider water once arrived on the land, as if it were an unvarying gift of God in heaven. Everyone knows full well that weather changes, but meteorologists know that climate too can change. Since at least the early 1970s, research on the global warming induced by human activities has shown that water resources, which fall from the sky, may well be distributed differently before the end of this 21st century. One of the reasons for this book is to remind readers that the water that we drink, the water needed to grow rice, wheat, and other crops, as well as to maintain forests and prairies, necessarily passes by way of the atmosphere that we humans are in the process of changing.

Another reason, related to the years I spent working in astrophysics, is to recall the history of water, which links our cells and those of all living creatures to the history of the universe. Water follows many paths on Earth, a unique planet as far as we know, covered with oceans and ice as well as deserts and green land, sheltering life on most of its surfaces. We certainly need to learn to manage our environment more intelligently, but we humans need also to dream and to satisfy our curiosity, our thirst for knowledge. Many years ago, on mission to the Haute-Provence Observatory in southern France to observe cool dwarf stars, I would curse the clouds. But who cannot be fascinated by the fantastic structures that sometimes take shape in the sky? Can you imagine the paintings of Ruisdael without the changing cloudscapes of Holland? In June 1979 I became a virtual astronaut, observing Earth from Meteosat[1] by working at a computer screen at the European Space Operations Center in Darmstadt, Germany. There I could admire the larger-scale structures of cloud systems extending over thousands of kilometers, while studying their variations with local time and their changes from day to day.

We cannot follow the full story of water, however, if we stay up in the clouds. In this book, I've had to trespass on geographers' territories, to fish in oceanographers' waters, glean in agronomists' fields, draw knowledge from the wells of the hydrologists. If I were to confine my discussions simply to my own area of specialization, it would interest precious few readers outside that area, and it would hardly help in understanding the broader issues. Of course, the problems of managing a water purification plant require the attention of specialized engineers, but one must also keep in mind where the water to be purified comes from, and where waste water will ultimately be discharged. Scientific research requires specialization and discipline, but excessively specialized scientific disciplines begin to resemble disciplinary cells if not solitary confinement, and I've no taste for that. Having traveled the planet Earth and some of its seas for over half a century, I claim the right to be interested in everything. That doesn't mean that I have the illusion of having an answer to every question. In Goethe's *Faust*, the old family servant Wagner says: "I've set myself to study fervently; I know it's very much, but I want to know it all."[2] Faust tends to make fun of him, but one could do worse.

The accidents of family history, life, and love have led me to make most of my scientific career in or near Paris, although I was born and grew up in New York City and studied and taught in the Boston area. I originally wrote this book in French, but seized the chance to translate it myself and adapt it for English-language readers, including examples of the wondrous ways of water from both Paris and New York and from other places on this planet that I know and love. I also have added some notes on the important contributions of French and other scientists to the widespread fields that I visit. I have taken into account some new scientific and political developments, although of course it is hopeless to try to make a printed book fully up-to-date. I hope, however, that I have succeeded in the challenging task of transforming the French-language structure into living English and not the constipated conference English that has become an international *lingua franca*.

One more point: some may find that my criticism of environmentalists or "ecologists" goes too far. Nonetheless, I believe I have a perfect right and indeed the duty to attack the exaggerations and at times completely wrong assertions that are far too current among environmental movements, no matter how good their intentions. Defense of the environment is too important to be left to advertising copywriters and pulpit-thumpers. Worse, those who cry wolf, when the threat is from the weasel, end up by comforting complacency. And those who work for the special interests of the merchants of coal and oil have too easy a time highlighting the errors, exaggerations, and inconsistencies of many "green" propagandists.

All children love water; you too, I hope. Castles of cloud, invisible vapor, breaking waves and spray, the calm of the ocean depths, peaceful lakes, rushing brooks, and the crystalline majesty of glaciers, all water. *Bon voyage*!

ACKNOWLEDGMENTS

My thanks go to Jean-Claude Pecker, who directed my doctoral dissertation in astrophysics and has continued to be a source of inspiration to me for more than forty years, even after I left that field. He also found for me the exact quote from Voltaire in chapter 5. Thanks also to the students at Virginia Polytechnic Institute and State University, who listened to me patiently in the spring of 1998. Thanks to Arielle, who carefully read part of my (French) manuscript, and to Jean Delaite, who hunted down unclear or misleading paragraphs in the French version. I also thank Holly Hodder, formerly of Columbia University Press, now at Westview Press in Boulder, Colorado, for encouraging me to take on the preparation of the English-language version. Finally, many thanks to Roy E. Thomas of Columbia University Press, for his thorough checking of the English-language manuscript, his stimulating questions and comments, and his help in clearing up the muddier passages.

water from heaven

PROLOGUE

PRIMING THE PUMP

Water—for most of North America, *water*; *wasser, voda* in the north and east of Europe; *eau, agua, acqua, hydro, maya* in western Europe, Quebec, and around the Mediterranean and Latin America: whatever you call it, water has always been an important part of our world, whether landscape, seascape, or sky. In Genesis (1:1–3), water and wind exist before light. In the Hebrew, the word *ruakh* denotes the wind as well as the spirit,[1] the breath of God on the waters before He lets there be light and separates day from night. And on the second day of the Creation (Genesis 1:6–10), God separates the waters of the Earth (*mayim*) from the waters of the heavens (*shamayim*). All this is to insist on the overriding importance of water for life. Don't get me wrong: only benighted "fundamentalists" still take literally (usually in English translation) the Genesis story of creation written down for unschooled shepherds; unfortunately, the partisans of the "creation science" swindle still manage to influence some state legislatures. Nor do I count water as an element. For Aristotle, water was indeed an element along with earth, air, and fire; but we have known for two hundred years that the elements are something else. Mendeleyev's periodic table now contains over one hundred elements, each element or atom made of particles more elementary still.

No water, no life; indeed, living creatures are more than two-thirds water. And if they could not count on the return of the rains, how could the first farmers settle down on the farm? With irrigation, one of the first acts of civilization, humans begin the great adventure, striving for ever-increasing mastery of the environment. They've come a long way . . . But the adventure sometimes turns sour, when man's apparent mastery betrays him,[2] as in the salinization, several thousand years ago, of once rich farmland in the Middle East.[3]

Even the driest of plants are about 50 percent water, a godsend for travelers of the desert who can count on calming their thirst with a few drops from the stalk-reservoir of certain cacti. Water, support and stuff of life, forces our respect. Let me try to tell its story. Some would have us call our planet not "Earth" but "Ocean," in honor of the liquid vastness that covers over two-thirds of the globe. But an extraterrestrial observer might well call our planet "Cloud," for these collections of liquid water droplets and solid ice crystals hide as much as 60 percent of the ocean or land surfaces of "Earth."[4]

Humans must always have known (guessed?) that clouds are made of water, even those that do not deign to water their crops. And those who have seen ice or snow—including inhabitants of the Tropics who occasionally have to hide from hail—know that these flakes or more substantial projectiles are nothing but water in a solid form, easily changed to liquid by heating. And yet there seems to be something of a paradox in the notion that air can contain water, when for Aristotle, and no doubt for many people still today, these are two distinct elements. In the Bible (Ecclesiastes 1:7) it is written: "All the rivers flow to the sea, and the sea is not full; and the rivers continue to flow . . ." What feeds the rivers? Some imagined that there were deep subterranean channels by which water found its way from the ocean bottom back up to the springs, somehow removing the salt along the way; but this won't work. Unless you can accept that there is continuous destruction of water in the seas, and continuous creation of water in clouds, the only sensible solution is that the waters of the seas are transformed into waters of the skies, the water from heaven falling back to land and sea—what we call the water cycle. You might think that this was understood long ago. After all, the still (for alcohol, if not for distilling water) is an ancient invention, and Leonardo da Vinci had a good intuitive understanding of the water cycle as early as the fifteenth century.[5] But it took alchemists and geographers some time longer to work out a reasonably correct picture.

In 1580, Bernard Palissy[6] showed how the water emerging in springs (the water of Earth) depended on the supply of water falling from the sky. Still, nearly another century went by before the Flemish physician and chemist Jan Baptist Van Helmont (1577–1640) outlined the concept of a *gas* as an invisible state of substances that are *distinct* from air even while they are *in* the air. There is water in the air, not only in the visible form of fog, mist, and clouds, collections of liquid water droplets and solid ice crystals, but also in the invisible form of gas. Today, scientists give the name of *water vapor* to that gaseous state of water, not to be confused with the occasional use of the word "vapor" to describe a visible mist composed of extremely fine droplets of liquid water suspended in the air. For Aristotle, water was an elementary substance; for some alchemists in the Middle Ages, it could nonetheless be changed into "earth" or "air." Since the work of Cavendish and Lavoisier in the eighteenth century, chemists know that water can be decomposed or atomized into the substances that we now know under the names hydrogen and oxygen. The name *hydrogen* (generated by water) was given (in French as *hydrogène*) by Guyton de Morveau, Lavoisier, Berthollet, and Fourcroy, the scientists who invented the modern nomenclature of chemistry in Paris, in 1787. Practically the same word is used in other Romance languages and in English, and also in Russian (becoming *vodorod*) and in German, where it becomes *wasserstoff*—the stuff of water.

THE STATES OF WATER

Atmospheric water vapor (i.e., water in the gaseous state) consists of H_2O molecules, in greater or lesser number, moving freely among the other molecules that make up air (about 99% being nitrogen and oxygen molecules). The formula for water, H_2O, goes back to the work of the French chemist Gay-Lussac[7] in 1805, and it means that each water molecule is made up of two hydrogen atoms and one oxygen atom. The formula could be written even better as H:O:H, to show that each hydrogen atom has its own link with the oxygen atom. The angle made by these two links is close to 105 degrees, and the hydrogen atoms are separated from the center of the oxygen atom by a distance of 0.096 nm (1 nm = 1 nanometer = one billionth of a meter, a meter being slightly over 39 inches). This folded structure, a sort of squashed letter V, can spin and vibrate, and because of this it can interact very strongly with infrared (heat) radiation with wavelengths from one to several hundreds of micrometers (1 micrometer or micron = 1 μm = one millionth of a meter; visible light radiation has shorter wavelengths in the range 0.4 to 0.7 μm). As a result, even though water vapor is a very small fraction of the atmosphere (less than 1%, even in the humid Tropics), it absorbs and sends back down a large part of the infrared radiation emitted by the Earth's surface. Water vapor thus dominates what is called the natural greenhouse effect, about which much more later.

On the Earth's surface, with temperatures in the range from 0 to 100° Centigrade (32 to 212° Fahrenheit, and seldom higher than 50°C = 122°F), we find water mainly in its liquid state. Over some areas of Earth, the temperature falls below 0°C (32°F) all or part of the time (but seldom below −60°C = −76°F), and then water is found mostly in its solid state as snow or ice. The English used to say "a pint's a pound, the world around," but nowadays most of the world has adopted the metric system, invented shortly after the French Revolution. A liter (1,000 cu. cm, about 1.06 U.S. liquid quart) of water is a kilogram (1 kg = 1,000 grams, about 2.2 pounds). Indeed, a gram is *by definition* the mass of a cubic centimeter of water.[8] Thus the specific mass or density of water is a gram per cubic centimeter, a kilogram per liter, a (metric) ton (1,000 kg, about 1.1 U.S. tons) per cubic meter. And as for the unit of distance, the meter was defined in terms of the circumference of the Earth, taken to be exactly 40 million meters or 40,000 kilometers (km).[9]

In liquid water, the water molecules are closely packed: their average separation is 0.27 nm, less than twice the size of the molecules themselves. The density of liquid water is fairly constant, as is the average distance between molecules. However, the molecules retain a certain freedom of movement relative to one another, and this is what makes liquid water liquid, fluid, quite different from the solid state in which the molecules are fixed in a more or less well-defined structure. I won't go into the quite complicated details of the physics of the liquid state.[10] Suffice it to say that with small quantities of liquid water, surface forces (surface tension) can be quite large compared with the force of gravity, and this is what makes it possible for water droplets to survive and to remain suspended (as mist or clouds) in the atmosphere, sometimes combining with one another to form bigger and bigger drops until they fall under their own weight as rain. Very large water drops can survive only in weightless conditions, as in a space station.

Some Remarkable Properties

Most of the time, water expands when it is warmed (for example, from 50 to 100°F) and contracts when it is cooled, just as do many other substances. This will by itself lead to a rise of sea level as the oceans warm in the coming century. But liquid water has a peculiar behavior: it has its maximum density at a temperature of +4°C (39.2°F). When temperatures drop below this level, water expands until it freezes. Although the density differences are small, they lead to the result that for near-freezing temperatures, the coldest water (below 4°C) has expanded, becoming less dense and floating near the surface, finally expanding still more as it freezes to form floating ice. The layer of ice at the surface of a lake protects the deeper layers, whose temperature will drop below 4°C only when all the higher layers have frozen solid.

Water also plays a special role as solvent, in other words as a medium containing other substances, sometimes in the form of molecules, sometimes broken up into atoms or ions, more or less uniformly distributed among the water molecules. For example, table salt, sodium chloride, its formula $NaCl$, has a cubical structure in its solid crystalline state, but this structure disappears when the salt is dissolved in water—in its place, individual $NaCl$ molecules, or else Na^+ and Cl^- ions paired up with the OH^- and H_3O^+ ions that can be formed from two water (H:O:H) molecules. The same is true for many other "salts," and for much bigger and more complicated molecules that play different specialized roles in the processes of living matter. The water in our blood enables it to carry the nutrients necessary for the different cells and organs of our body, and to transport waste products to the appropriate organs of elimination, such as the kidneys.

Gases also can be dissolved in water. At "room temperature" of 20°C (68°F), a liter of water can contain 19 milliliters (cubic centimeters) of air: and fish "breathe" by extracting oxygen from air dissolved in the water. And because air is dissolved in our blood, the nitrogen (N_2) molecules that make up 78 percent of air cause problems for deep-sea divers who come up to the surface too quickly, forming bubbles in the condition called "the bends." As for carbon dioxide (CO_2), the total amount dissolved in the oceans is far larger than what is present in the gaseous state in the atmosphere. The gases and solids dissolved in a body of water will of course affect its physical and chemical properties. The exact density of seawater depends on its salinity (between 17 and 33 grams of salt for a kilogram of water), so that some water masses sink to the bottom, while others tend to spread out at the surface. At the underwater threshold of the Straits of Gibraltar, a sort of super-Niagara of extremely salty water flows out from the Mediterranean and sinks into deeper layers of the Atlantic Ocean, while less salty water flows in at the surface from the Atlantic.[11]

In the solid state constituted by the different forms of snow and ice, the atoms of hydrogen and oxygen that make up water are arranged in well-ordered crystal structures. On our planet, ice is found in the form called *Ih* by specialists, a nearly regular framework of tetrahedrons 0.37 nm high, piled up in layers with their bases arranged on equilateral triangles of 0.45 nm on a side. Individual water molecules can be identified with practically the same size and shape as in the liquid state, but the difference is that each H:O:H structure is linked to its neighbors by the protons (the nuclei of the hydrogen atoms) that jump back and forth between the neighboring locations at 0.096 nm from oxygen atoms. Thus a pair of protons is shared by two oxygen atoms, but each of the oxygen atoms is associated with two pairs of protons. As a result, the whole structure takes up a bit more room than in the liquid state; and that explains why ice has a lower density than water. The triangular bases of ice *Ih* lend themselves to forming hexagonal columns and plates and more complicated and beautiful six-branched structures in the enormous variety of snowflakes, but these are not the only possible forms of ice. Ten other forms have been produced in the laboratory (ice *Ic*, with cubical rather than hexagonal symmetry, ices *II–IX*), but they exist only at very high pressures and very low temperatures. Perhaps some of these forms exist naturally under the surfaces of Ganymede and Callisto, two of the four moons of Jupiter discovered by Galileo.

WATER'S MANY CYCLES

We speak of *the* water cycle as if there were only one, but in fact there are many: some circuits are completed in a few days; but in others it can take months, years, millennia, even millions of years to go the course. Over tropical seas, strong sunlight and steady trade winds combine to evaporate thousands of tons of water per square mile every day. Some of that water may rain back on the sea only a few hours later, but on average the water remains a week or more in the atmosphere before returning to the surface.

Water that stays in the atmosphere for ten or more days can go a long way, reaching areas far inland. If it falls on mountain peaks in the form of snow, it may take months before it melts and returns to the sea by way of streams and rivers. In glaciers, centuries may go by before the snow-turned-to-ice melts, and water that has seeped underground may spend thousands of years there before reemerging.

Elsewhere, water soaked up by the ground is tapped by life—essentially plant life—and this is precisely what limits runoff, i.e., the direct return of liquid water to the oceans by way of brooks, streams, and rivers. As everyone knows, plants need water. They use it in three ways. Part of the water pumped from the soil through the root system returns rapidly to the atmosphere by way of the *stomata* (tiny openings in the leaves), a process called evapotranspiration that keeps the plant from overheating. Another share is taken up directly in the plant as water, without any modification. And finally, a third share is transformed into *organic* matter—complex molecules made of carbon, hydrogen, oxygen, etc.—by way of the chemical process known as *photosynthesis*, activated by sunlight. Months, years, or in the case of certain trees, even centuries may go by before all the hydrogen and oxygen atoms of that share return as water to the environment, following oxidation of the organic matter.

Cycling back and forth between the oceans, the atmosphere, the land, and the living matter of the biosphere, water also circulates within the ocean. Most people know about the Gulf Stream, crossing the Atlantic in ten or so years. But ocean currents have a third dimension, with cold salty brine plunging to the depths of Davy Jones, returning to the surface centuries later on the other side of the world. And some of the snow that falls in Antarctica may remain frozen for millions of years, preserving the memories of past climates in the ice.

PART I

Water in the Universe
from the Big Bang to the Appearance of Man

CHAPTER 1

BEGINNINGS

From Hydrogen to Oxygen, and All the Rest

Where—or what—does water come from? Lavoisier and Cavendish knew how to make hydrogen from water at the end of the eighteenth century; but according to the formula H_2O that Gay-Lussac first wrote down in 1805, water is a molecular edifice made up of the *elements* hydrogen and oxygen. We know that these elements exist elsewhere in the universe, and hydrogen is by far the most abundant element in the Sun and stars, indeed nearly everywhere. As for oxygen, if we count its abundance as expressed in number of atoms (rather than in mass), it is the third most abundant element in the universe, coming after hydrogen and helium. Even so, the amount of oxygen is minute on the cosmic scale, at most one thousandth—a tenth of a percent—that of hydrogen. How can we know this? In 1835 the French positivist philosopher Auguste Comte (1798–1857) solemnly asserted that the composition of heavenly bodies was positively unknowable, but by the end of the nineteenth century, the advances of astrophysics and of spectroscopy proved him wrong. The scrutiny of the spectra of the Sun (as in a rainbow) or of stars—in other words the study of the detailed distribution of energy with wavelength in sunlight—or starlight or other radiation, reveals dark and bright spectral lines and bands. These features are fingerprints of the presence of specific atoms or molecules. From the relative strengths of these lines and bands, we can determine the chemical composition and the physical conditions at the surface or in the outer layers of the atmospheres of the heavenly bodies, without ever getting a sample of the material in our laboratories.

Hydrogen—the stuff of water—is by its abundance the stuff of which the universe is made; and oxygen, the other component of water, turns out to be the least rare of the other elements, with the sole exception of chemically inert helium. Admittedly, the development of nuclear physics has shown that atoms are not indivisible, thus contradicting the very meaning of the Greek word *atom*. The elements are made of still more elementary particles. Since the 1920s and '30s, we know that: (1) an atom is made up of a nucleus containing nearly all of its mass; (2) the nucleus is made up of particles with a positive electric charge, called *protons*, and usually also particles with the same mass but *no* electric charge, *neutrons*; (3) negatively charged particles of extremely small mass, the *electrons*, are found in orbits around the nucleus, at very great distances compared with its size; (4) only certain electron orbits are allowed following the rules of quantum mechanics; and finally (5) spectral lines correspond to the jumps of the electron from one permitted orbit to another. The atom as a whole is electrically neutral, but it can lose—or sometimes capture—one or more electrons, and it then has a net electrical charge, becoming what is called an *ion*. Different atoms can combine, sharing or exchanging their electrons and making up larger edifices—*molecules*—and these can vibrate and rotate, again not in any which way but rather following the strict rules of quantum mechanics. As a result, these states of vibration and rotation also leave their fingerprints in the spectrum. In chemical reactions, atoms combine or separate and molecules are formed or broken apart, and in all this they release or absorb heat and emit or absorb radiation, as their states of agitation, vibration, and rotation change.

Since the discovery of radioactivity by Henri Becquerel in 1896, since the work of Marie and Pierre Curie and all that followed, we know that transmutation of the elements is possible. Atomic nuclei can be transformed by what is called *radioactivity* (spontaneous emission by the nucleus of particles and/or radiation, changing the numbers of protons and/or neutrons), by nuclear *fission* ("splitting the atom," dividing a nucleus into two nuclei each of smaller charge and mass), by nuclear *fusion* (combining two or more nuclei to form a nucleus of substantially greater mass and/or stronger charge), and by other reactions between colliding nuclei. Since the 1930s, physicists studying such high-energy collisions have discovered a multitude of so-called elementary particles, revealing complex patterns interpreted in terms of an extremely abstract theoretical framework. The details of the weak and strong interactions of these different particles are the matter of the new physics developed in the latter half of the twentieth century. I won't go into that here; for our explorations, it will be enough to consider protons, neutrons, and electrons as the building blocks of matter. However, observation has confirmed the theoretical predictions of the existence of *antimatter*, made of antiprotons (same mass as the proton, but negatively charged) and positrons (anti-electrons, of the same extremely low mass as the electron, but with positive charge). Inside a nucleus, a proton can change into a neutron by emission of a positron, and a neutron that has left a nucleus will disintegrate, on average within eighteen minutes, into a proton and an electron; on the other hand, when a positron and an electron collide, they are mutually destroyed in a burst of extremely energetic gamma radiation. In our universe of matter, antimatter has an extremely short lifetime.

Hydrogen, the stuff of the universe, is indeed the most elementary element that exists: in its most common form, the atom is formed by a nucleus reduced to a single proton, with a single electron orbiting around it. In 1929 the American astronomer Edwin Hubble (1889–1953) discovered that all distant galaxies (the word *galaxy* comes from the Greek for *milk*)—assemblages of stars (possibly accompanied by planets), gas, and dust, analogous to our own Milky Way—appear to be moving away from us at speeds proportional to their distance. Interpreted as an expansion of the universe, this leads to the conclusion that all of the observable universe was concentrated in a compact volume some fifteen billion years ago. The Belgian cosmologist and priest Georges Lemaître (1894–1966) called this fantastically hot and dense volume the "cosmic egg," but since the 1950s it most often goes under the name of the "Big Bang." Fred Hoyle[1] originally thought up the name to make fun of the concept, but it seems to have stuck. Today, most cosmologists believe that there really was a Big Bang, not that they necessarily take it as confirmation of the (relatively confused) cosmogony of the book of Genesis. Working together with theoretical and experimental high-energy physicists, they seek to describe and understand the first moments—the first three minutes,[2] the first second, perhaps the first billionth of a second—of the universe. Passing over all the details and debates, I shall summarize: very rapidly, in less than a second, a universe of radiation, matter, and antimatter is formed. For about a year, this primeval fireball cools as it expands, the temperature falling to ten thousand billion degrees, one billion degrees, ten million degrees. During these periods, *nearly all* the matter and antimatter particles of the universe are annihilated, yielding energetic gamma-ray photons (electromagnetic radiation). Somewhere in this chaos, protons and electrons survive, an ordinary matter majority that gives rise to our universe. Another hundred thousand years of expansion go by, the temperature falls to 3,000 degrees Kelvin (K) (about 5,000°F), and protons and electrons link up to form hydrogen atoms. This is called the *recombination* period, even though it was the first time that such neutral atoms came to exist. For ever after, most photons will go their own way independently of matter. We still can "see" today the past glory of the young universe in the form of radio waves coming from all directions, with wavelengths a millimeter or so. This "cosmological black body radiation" has such long wavelengths, corresponding to a temperature of only 2.7°K (−454°F), because it is so strongly "redshifted" by the expansion of the universe. Slight irregularities in this background radiation give astronomers the key to understanding the formation of galaxies and clusters of galaxies that dominate today's universe.

We have, in the hydrogen of the water of our bodies, a representative product of the Big Bang, the process of formation of our matter universe. But what about the oxygen in the water? Where does it come from? During the first hour of expansion of the primeval fireball, the very high densities and temperatures of millions of degrees Kelvin must have led to fusion of some of the hydrogen nuclei (protons), producing the quite rare heavy hydrogen isotopes (deuterium: one proton, one neutron; tritium: one proton, two neutrons) still observed today. The chain of thermonuclear fusion reactions leads to substantial quantities of helium nuclei (two protons, two neutrons; and a fairly rare isotope composed of two protons and one neutron). The great Russian chemist Dimitri Ivanovich Mendeleyev (1834–1907) did not include helium in his first periodic table of the elements, published in 1869. Discovered only later, by its "signature" in the spectrum of the Sun (which is why it was called helium), this element is chemically inert. The rules of quantum mechanics allow (oblige) it to hold jealously on to its two electrons, virtually never sharing them with other atoms to form molecules. The helium nuclei, also called alpha-particles, can take part in nuclear reactions (reactions between nuclei); but the expansion of the primeval fireball proceeded too rapidly for significant numbers of elements heavier than helium to be produced. Again, we have to ask the question: how was the oxygen of the water, how were all the other elements found in stars, planets, and living matter produced?

When the universe was still young, but after the expansion and the cooling had led to a halt in thermonuclear reactions, small irregularities in the distribution of the matter of the universe—mostly hydrogen, but with some helium (less than 8% by number of nuclei, but about 30% of the mass)—begin to grow, with matter collecting in denser blobs here and there, rarefying elsewhere. Galaxies and clusters of galaxies appear. Gravitational attraction then leads some of the accumulations of matter to contract still further. The interactions between gravitational attraction, rotation, and magnetic fields can be quite complicated, but again, without entering into details, I can write that some of these contracting collections of matter rapidly form a first generation of stars, mostly tens of times more massive than our Sun.

If indeed the mutual gravitational attraction overcomes the rotational resistance (centrifugal force) opposed to contraction, the mass of gas continues to contract, becoming denser and denser. Collisions between atoms become more and more frequent, more and more violent, and the gas heats up. Finally, temperatures inside some of these collections of hydrogen (and helium) become so high (a million degrees Kelvin, 1.8 million degrees Fahrenheit) that they trigger resumption of the thermonuclear fusion reactions transforming hydrogen to helium. In the infant universe, expansion and overall cooling go so fast that thermonuclear reactions come to a halt after one year; and the expansion continues today. In the meantime, within this ever more rarefied expanding universe, there exist concentrations of matter held together by the attraction of gravity. For sufficiently large masses, densities and temperatures near the center reach values high enough to maintain thermonuclear reactions. The blob of matter becomes a *star*, and the fusion of hydrogen to form helium can go on for millions, billions, and even tens of billions of years, the accompanying energy output accounting for the star's brightness. For very massive stars (tens of solar masses), the hydrogen in the central core runs out in a few million years;

gravitational contraction then resumes, densities and temperatures rise still more, until they reach the point at which fusion of the helium nuclei—the alpha-particles—sets in. It turns out that according to the quantum rules of nuclear physics, the first element that can then be produced in abundance is carbon (six protons and six neutrons in the most common isotope), as a result of what is called the triple-alpha reaction.

Life, based on carbon, thus owes its very existence to the fusion of helium nuclei that took place in one or more generations of extremely massive and hot stars, early in the history of the universe. Further reactions follow: adding still another alpha-particle to a carbon nucleus (under still hotter conditions) yields the oxygen nucleus (eight protons, eight neutrons in the common isotope) needed to form water. And after oxygen-16, adding further alpha-particles yields neon-20, magnesium-24, silicon-28, and so forth up to titanium-48, if the central temperature gets as high as three billion degrees (only the case for a very massive star). Other more complicated reactions account for the other elements in nature, in particular iron-56. These reactions proceed more and more rapidly, and these most massive stars end their lives in less than ten million years, in a violent explosion, a supernova, scattering part of the stellar matter enriched in heavy elements.

Today, massive stars of the first generation, composed solely of hydrogen and helium, are nowhere to be seen. Consuming their stock of hydrogen and helium at dizzying speed, their lives lasted but an instant compared with the age of the universe (several million versus fifteen billion years) and ended in supernova explosions a long time ago. Perhaps some of them contributed to the feeble glow of remote galaxies, only detected with great difficulty today after billions of light-years of transit, but there is no way to identify them. The supernovae observed today (or a thousand years ago) do indeed enrich the interstellar medium with the products of the stellar "furnace," but these are explosions of stars only several million years old, already containing elements heavier than hydrogen and helium at their birth. Could there still be stars of the first generation close by? They would have to be of quite small mass, with lifetimes of tens of billions of years. Astronomers have indeed observed certain populations of stars whose spectra reveal very low abundance of heavy elements such as metals. Inasmuch as the spectrum provides direct information on the composition of only the outermost layers of the star, it is possible that these have been "contaminated" by the enriched matter disseminated between the stars by explosions of more massive but shorter-lived later-generation stars. However, it is possible to estimate the proportion of heavy elements from the overall properties of a star: mass, luminosity, radius. It turns out that most stars observed today contain a small but significant proportion (a few thousandths) of elements heavier than helium: carbon, nitrogen, oxygen, sulfur, silicon, iron, etc. These stars of the mature universe contain the products of prodigal massive stars of the first generation. Having produced heavy nuclei in their cores, they scattered them, together with leftover hydrogen and helium, in a spectacular final act. The stuff of water—not the hydrogen but the oxygen—just like the stuff of planets and of life—iron, silicon, magnesium, aluminum, carbon, nitrogen, sulfur, phosphorus, and so on and so on—comes from stardust, the stuff of those splendid stars that shone when the universe was young.

WATER IN THE INTERSTELLAR MEDIUM

So little matter exists between the stars—less than an atom per cubic centimeter—that one might be tempted to write the interstellar *void* rather than *medium*. But here and there we encounter "clouds" hundreds or thousands of times denser. Explosions of stars violently eject matter—hydrogen, helium, and the heavier elements—into a nearly empty cold medium. Around other stars, in particular around "red giants," enrichment proceeds less dramatically. During the red giant life stage of certain stars, helium is consumed after the exhaustion of hydrogen in the dense contracted central core, but the outer envelope of the star expands enormously like a balloon. The Sun too will one day—more than four billion years from now—become such a red giant, engulfing the planet Earth and burning it to a crisp. In these extended envelopes, temperatures are "low" enough (less than 3,000 K or 5,000°F) for molecules to form, not only such tightly bound molecules such as carbon monoxide (CO), whose spectrum is observed in sunspots, but also water molecules. Solid particles—ice, grains of silicates or carbon compounds—can also be formed and can survive. At the "surface" (really in the outer layers) of these stars, the weak gravitational attraction of the distant center and the strong flux of radiation coming from the center give rise to a "stellar wind," a gentle expulsion of matter toward the interstellar medium. Some of the stardust in our bones comes from those dusty breezes as well as from the more spectacular blasts.

Some of the space between the stars appears to be filled by bubbles of hot but extremely rarefied gas, but elsewhere colder and much denser condensations are found. Stars continue to be born today, and in some cases astronomers observe the birth process itself, or if not, the extremely young newborn stars, generally within clouds of gas and dust containing thousands of particles per cubic centimeter, sometimes as cold as 20 K ($-423°F$). How does matter go from these highly dilute bubbles of hot gas to the denser and darker cold clumps of complex molecules and solid particles? Astrophysicists are far from understanding all the details. Nevertheless, such clouds exist: solid grains weaken and redden the light of distant stars, and they leave fuzzy spectral signatures that seem to be those of silicates, of ice, of graphite, and perhaps also of fullerenes (a soccer-ball-like structure of 60 carbon atoms, named after Bucky Fuller).[3] Stellar grains may also act as condensation or freezing nuclei, surrounding themselves with a mantle of water ice. Once solid particles (grains) form inside an interstellar cloud, it can cool off very effectively, and still more complex molecules can form; but we still do not understand completely how the first solid grains form. Spectroscopy and radio astronomy confirm the existence of molecules in these clouds: not only hydrogen (H_2), water (H_2O), methane (CH_4), and ammonia (NH_3) but also more complex organic molecules (alcohols, ethers) made up of ten or more atoms.

Some scientists have called such matter "prebiotic" because of the presence of organic molecules, and believe that it is indeed the birthplace of life. But are terrestrial conditions so unfavorable that we have to look elsewhere? In any case, these hypotheses are no answer; they just put the problem of origin of life further off. The raw materials of life, the elements heavier than helium, in particular carbon and oxygen, are forged in the furnaces of the stars. From the universal matter hydrogen and star-produced oxygen, Mother Nature has little difficulty fabricating water. Let's go on with its story.

CHAPTER 2

THE CHURNING OF THE EARTH

SAVING THE WATER AT THE BIRTH OF THE PLANETS

Are we alone in the universe? Today, our planet Earth, with atmosphere, clouds, and oceans, surrounding a mostly solid mass made of heavy elements, appears to be the sole abode of life among the planets orbiting our Sun. On dry and frigid Mars, we can discern some vestiges of a gentler climate, of a liquid age when water flowed on the planet's surface. A few peculiar meteorites, fragments from Mars found on Earth, harbor features that some researchers have identified as fossil Martian bacteria. Others disagree, and the question is not settled.[1] Could some such microbes still survive in hibernation somewhere in the long frozen soil of the red planet, awaiting the return of better days? Could human colonists, engaging in an activity of "terraforming,"[2] render the Martian climate less hostile to life? All these questions and more are raised as we terrestrials wonder how far and how fast to push the exploration that began when the *Viking*-1 and -2 spacecraft reached Mars in 1976. Increasingly sophisticated robot equipment, with no requirement for a round-trip ticket, has sent us a wealth of data in recent years, answering some questions and raising new ones.[3]

And elsewhere? Dry Venus, with its thick atmosphere (more like a gaseous ocean) of carbon dioxide, has temperatures at the surface that are infernal, and the planet hardly seems hospitable to life. Things are no better on Mercury, with no atmosphere and temperatures oscillating regularly between infernal heat and deadly cold. And the giant planets are but failed stars, no place for life anything like what we know. What about their satellites? Europa, a moon of Jupiter, and Titan, a moon of Saturn, appear to be covered with ice, perhaps floating on oceans despite the extreme cold.[4] Fascinating questions, but we might as well wait for the data that the Huygens-Cassini space probe will send back.

Ever since Copernicus put the Sun, rather than the Earth, in its central position, we call our planetary system a solar system. Is it unique? Along with the Copernican Revolution came a realization of the vastness of the universe, a realization that for stars to be visible at all at their enormous distances, they must be suns; and by the same token, the Sun is just an ordinary star. Once this is accepted, it is natural to ask whether there could not be many planetary systems and indeed inhabited planets, around other stars in the universe.[5] For our world of earth and water to exist, stars had to be born and die, and our Sun and its retinue of planets be formed from the debris of past generations of stars.

A few thousand stars are visible to the naked eye, and astronomers count thousands of billions of them. But, at best, only a dozen or so planetary systems have been conclusively identified so far. It is nevertheless believed that planetary systems abound in the universe, that the formation of planets often goes along with the process of star formation. In order for a cloud of matter to become a star, it has to contract, becoming denser and hotter. However, while the mutual gravitational attraction of the matter in the cloud tends to pull it together, the cloud's rotation will accelerate at the same time. This results from the law of conservation of angular momentum: anything that spins continues to spin unless its spin is slowed down by increasing another body's spin, and the "angular momentum" that measures the spin depends on the spin rate and on the distance from the axis of rotation. Just as a skater will spin more quickly by bringing her arms in toward her body, so the contracting cloud will rotate faster and faster. For the same reason, a comet in an extremely elongated elliptical orbit around the Sun moves much more quickly when it is close to perihelion than when it out beyond Jupiter's orbit. Everything in the universe is rotating: galaxies, nebulae, stars, planets and their satellites. It is hard to believe that all stars are born from the very few clouds that are not rotating. However, as a contracting cloud of gas and dust spins more and more quickly, it is distorted by centrifugal force and ends up losing mass in its equatorial plane in a sort of pre-stellar wind. Can it ever contract enough for the temperature and density in its central core to reach the "ignition point" for thermonuclear reactions, to start it off on the life of a true star?

Despite these difficulties, stars shine. How do they rid themselves of their angular momentum? For many of them, the answer is that the cloud has split up to form a binary or multiple stellar system: the angular momentum is still there, but it's been concentrated in the motion of the two or more stars around each other. Some single stars—always very young—spin very quickly on their axes, but many single stars rotate very slowly (in 25 to 27 days in the case of our Sun). However, although the angular momentum of the Sun itself is quite small, that of the solar system as a whole is not. Although the planets contain only 0.13 percent of the mass of the Sun, their orbital motions account for nearly all the angular momentum of the solar system, 15,000 times more than the rotation of the Sun alone on its axis. In the process of formation of the Sun and its retinue of small and giant planets, nearly all the angular momentum was somehow transferred from the central condensation or "proto-sun" to the surrounding "protoplanetary" cloud. This transfer process depends on the existence of magnetic lines of force that accelerate the distant cloud's rotation at the expense of the spin of the central condensation. Not all the details have been worked out, but it's hard to see an alternative to this scenario. Within the flattened spinning cloud, protoplanetary condensations must have formed, becoming the planets and their satellites quite quickly (in less than 100 million years). All the planets revolve around the Sun in the same direction and in nearly the same plane (the ecliptic) close to the equatorial plane of the rotation of the Sun itself, and nearly all the planets rotate on their own axes; and for those that have satellites, those satellites also revolve around them in the same direction. Such a scenario is not likely to be peculiar to the solar system; it accounts for the multitude of apparently single slowly spinning stars. And while only a few planets have been identified with certainty outside the solar system, astronomers have shown, mostly using infrared observations, that many stars are surrounded by disks of gas and dust. Far from being an exception, planetary systems may be quite common.

Interstellar, protostellar, and proto-planetary-system clouds contain hydrogen, helium, and other elements in gaseous form as atoms, ions, and molecules. The H_2O molecule must be there. These clouds also contain solid grains of different sorts, sometimes coated with sheathes of ices, often water ice. As gravitational contraction proceeds, other grains may condense and trap more water. However, in the gestation that preceded the birth of our Sun, the gradual warming of the protosolar cloud leading to the firing up of the thermonuclear furnace also warmed up the protoplanetary cloud. Far from the center, temperatures remained low, so that even grains composed of volatile substances such as water, methane, and ammonia survived in solid form. But closer to the central source of warmth, as temperatures rose above 1,000 or even 1,500 degrees K (about 2,400°F), water ice evaporated, and only more or less refractory grains survived together with silicates and metallic iron. In between, other minerals could be found, in particular hydrated minerals in which water molecules (in fact broken up into hydrogen atoms and OH radicals) were trapped in the mineral crystal structure. At this point, a strong "wind" from the young Sun swept through the flattened cloud, in its central part sweeping away nearly everything not solid (or not attached to a solid grain). In the inner part of the solar system, where today orbit Mercury, Venus, the Earth-Moon couple, and Mars, nearly all of the mass, practically all the hydrogen and helium, was swept away and lost for the formation of these planets. The helium found on Earth, extremely valuable for safely flying balloons and dirigibles, is not a residue of the Big Bang but rather a product of the radioactive decay of uranium formed in earlier generations of stars. The argon-40 isotope, nearly 1 percent of air, comes from the decay of radioactive potassium. The other argon isotopes, like the other rare gases (neon, krypton, xenon) are extremely rare (both in an absolute sense and compared with their abundance in the Sun), as a result of the violent protosolar wind that cleaned out the inner protoplanetary clouds. These inert gases could not form compounds with other elements or be bound up in minerals.

Farther out, between the orbits of Mars and Jupiter, hydrogen and helium were also swept away. Here, source of most meteorites, orbit the asteroids, possibly leftovers of a shattered planet or maybe of one that never formed. Farther away still, the protosolar wind was weaker, and the protoplanetary condensations that later became Jupiter, Saturn, Uranus, and Neptune kept most of their gaseous hydrogen and helium; indeed their composition is pretty close to that of the Sun. So this, painted in broad strokes, is how our solar system was formed. There are of course plenty of features that still need to be filled in, as astronomers try to explain in detail the differences in chemical and isotopic composition among the different planets, comets, and meteorites.

PRIMORDIAL WATERS—FROM INSIDE THE EARTH OR FROM OUTSIDE?

Buried in the Earth's crust, miners can extract—blessing or curse?—uranium and other radioactive elements made in a supernova explosion less than five billion years ago. Old Mother Earth is still young enough to have kept some traces of this event. After ten billion years, the uranium would be much rarer, half of it decaying to lead every 4.5 billion years. Whether from meteorites fallen to Earth, or rocks from the Moon or Mars, the isotope ratios all point to condensation of the solid bodies of the solar system between four and five billion years ago. Which ratios? Generally, the ratio of the abundance of a radioactive isotope to that of the stable isotope resulting from its decay (for example, the ratios uranium-238/lead-206, uranium-236/lead-207, potassium-40/argon-40, etc.). Earth's childhood having been particularly tumultuous, few rocks older than three billion years are to be found today. Nevertheless, the specialists are quite convinced that, give or take a few hundred million years, the solar system was formed 4.6 billion years ago.

At the very beginning of this process, the condensation destined to become Earth was surrounded by a cloud consisting mainly of hydrogen and helium, with some water vapor as well. However, this primary atmosphere of the Earth must have disappeared in the course of the clean sweep by the young Sun's radiation and wind. But where then does the water of the Earth come from? Did heavenly waters arrive on the surface of an already formed planet? For a billion years following the birth of the Earth, its surface, like the surfaces of the Moon, Mars, and Mercury, was intensely bombarded by meteorites, some of which contained water.[6] The marks of these impacts have, for the most part, long been erased by erosion on Earth, to a lesser extent also on Mars, but they are still clear on the pockmarked surfaces of the Moon and Mercury. There are over a billion billion tons of water on Earth, and it is difficult to calculate exactly how many tons of impacting meteorites were necessary to supply it. The water needed to fill the oceans must have come with enough debris to make a layer a hundred kilometers thick, much thicker than the Earth's crust as we know it today.

Comets may be another source. Far beyond the orbits of the giant planets, small clumps of matter must exist very nearly undisturbed since the beginning: a little dust, surrounded by water and other ices. These "dirty snowballs," as Harvard astronomer Fred Whipple called them, only reveal themselves as comets after perturbation of their motions by a passing star, sending them into extremely elongated elliptical orbits that take them into the inner part of the solar system. Closer and closer to the Sun, warmer and warmer, some of the ices evaporate, and a cloud of gas and dust escapes from the semisolid nucleus of the comet. Swept away from the Sun by the pressure of its radiation and of its wind of particles, the cloud becomes a *coma* (hair) and sometimes a splendid tail shining in the skies for several weeks.[7] Today, fortunately, major comet impacts on the Earth are rare—at least we'd like to think so. And earlier?

Are catastrophic comet collisions needed to carry water to Earth? There has been a long-running scientific controversy about the possibility that "micro-comets" might constitute a continuing supply of water for the Earth, even today. In 1981, Iowa University physicist Louis Frank was studying measurements of ultraviolet radiation from the night side of the Earth using an instrument on board the NASA Dynamics Explorer 1 satellite, and he identified spectral emission lines due to upper atmosphere nitrogen and oxygen atoms, excited by the particles of the solar wind.[8] On the images, Frank found many black spots, which he attributed to comets of very low mass (10 to 30 tons). Frank's estimate was that twenty or so such micro-comets collided with the Earth's atmosphere every minute, bringing enough water to cover the Earth with several centimeters (or a few inches) extra every 20,000 years. If such a bombardment went on for three billion years, it would be enough to renew all the water in the oceans many times over. Following the launch in 1996 of NASA's *Polar* satellite, Frank also found ultraviolet emission lines due to the hydroxyl (OH) radical, and he argued that this confirms the idea that these objects contain water. However, no signs of such frequent impacts have been observed on the Moon or Mars. Neither are there signs of water enrichment in the atmospheres of the giant planets, even though they are much more powerful collectors of interplanetary debris than the Earth. Could the lines identified with OH, emitted at over a hundred kilometers altitude, be due to water molecules already present in the Earth's atmosphere? That too is hard to believe: the upper atmosphere is extremely dry, and water vapor seldom gets above the level of the tropopause (10–15 km altitude). Many specialists who have looked at the data believe that the black spots are instrumental artifacts and not due to atmospheric features, but Louis Frank and his colleagues have not conceded defeat. The puzzle remains.[9]

Water arrived from the skies, but the atmospheric water of the first ages of our planet must have been lost if not recycled in the Earth long ago. Water arrived as the first solid matter condensed to form the proto-Earth, it arrived during the accretion phase as the planet acquired its final mass, perhaps also during the following billion years. It arrived in solid or frozen granules and in the hydrated minerals that accumulated as the planet formed, perhaps also with the rain of comets and meteorites, the debris of asteroids. These granules are cold as they arrive, but the impact shock of agglomeration and accretion heats up the surface. At the same time, the (proto-)planetary surface absorbs solar radiation, converting it into heat, a process that necessarily continues today. Moreover, the process of consolidation of the infant Earth took place only a short time (maybe 300 million years) after the cloud contracting to form the solar system was enriched by a supernova explosion. The intense radioactivity of short-lived nuclei produced by such an explosion must have heated each of the proto-planets still more, not only at its surface but throughout its volume. Even today, over four and a half billion years later, the long-lived radioactive nuclei produced by the same supernova are responsible for much of the heat flux emerging from inside the Earth: about a tenth of a watt per square meter (that still adds up to 50 million megawatts for the whole planet), very little compared with the 240 watts per square meter absorbed from sunlight (120 billion megawatts for the planet). Most geophysicists believe that water, like the other components of the ancient secondary (not primary) atmosphere, emerged as a result of the "outgassing" of the outer solid layers of the Earth.

OUTGASSING AND DELUGE

Today, the oceans cover 71 percent of the planet's surface; their average depth is 3,700 meters (over 12,000 ft.), so that they contain 1.35 billion cubic kilometers or 1,350 billion billion liters of water. Enormous as such a figure is, it still represents only 0.02 percent of the total mass of the Earth.

The only way to get such a huge quantity out of the Earth's interior was to make it boil. Even though the proto-Earth started to form in the cold, Earth's early infancy took place in heat released by intense radioactivity over the following few hundred million years. Much of the abundant metallic iron of the first condensations must have melted, heated by the accretion of still more iron-rich material, and this molten iron must have flowed down toward the center of the Earth, forming a solid metallic inner core surrounded by liquid. Together, this iron and nickel core contains about a third of the mass of the Earth, with temperatures reaching 7,000 °C (12,632°F). The slow infiltration of liquid iron to the central region is in effect a particularly gentle sort of fall, converting gravitational energy into heat just as in the case of the violent fall of meteorites. The intense heat released inside the Earth at its formation, reinforced and trapped by the inner core's solidification, supplemented by the heat of radioactive decay, can escape only through the mantle, intermediate layers of less dense material that comprise two-thirds of the Earth's mass. Although not liquid, at temperatures of a few thousands of degrees Kelvin, the mantle is plastic (in the technical sense), i.e., it can flow, and the convective flow transfers heat outward. The temperature structure of the mantle and the strength of its convection depend on the temperature in the central core, the radioactive release of heat in the mantle, and the rate at which heat can be lost at the upper surface.

Under these conditions, certain minerals and elements tend to separate from others, ending up near the surface of the globe in a thin chemically different crust (up to 70 km or 43 mi. thick, only 0.4% of the mass of the Earth). This is the *lithosphere*, less dense but much more rigid than the mantle. Below it, in the mantle, heated minerals give up water, carbon dioxide (CO_2), and other gases that they contain, and these gases tend to rise. Here and there they find their way to the surface through cracks and vents in the rigid lithosphere, emerging from the smoking mouths of volcanoes. Such a process, no doubt many times more intense than today, must have given rise to the planet's "secondary" atmosphere.

The slow settling of iron, together with radioactivity, have kept the center of the Earth hot, even while temperatures at the surface have stayed in a reasonable range over most of the Earth's history. And yet the Sun was almost certainly fainter in the past, perhaps 40 percent fainter some four billion years ago. If the Earth's atmosphere then was the same as today's, the

Earth would have been completely frozen over.[10] But today's atmosphere, rich in oxygen but relatively poor in carbon dioxide, is a product of the evolution of land plants. At first, the atmosphere contained very little oxygen; on the contrary, it contained much more CO_2 and other gases. These gases contribute to what is known as the *greenhouse effect*, trapping thermal infrared radiation (invisible heat radiation, with wavelengths greater than four microns) in the lower layers of the atmosphere. By contrast, solar visible and near infrared radiation at shorter wavelengths passes quite easily through the atmosphere. The greenhouse gases, molecules made of three atoms or more (H_2O, CO_2, NH_3, CH_4, etc.), are relatively rare in the atmosphere, but they are effective absorbers of infrared radiation. On the contrary, the nitrogen (N_2) and oxygen (O_2) molecules have structures far too simple for them to interfere very much either with sunlight or with the thermal infrared radiation emitted by the Earth. Three billion years ago, the natural atmospheric greenhouse effect must have been much stronger than today, keeping the Earth's surface warm despite the fainter Sun of those days.

As outgassing of the Earth's mantle proceeds, spouting more and more water vapor into the atmosphere, the greenhouse effect increasingly warms the surface and lower atmosphere. However, when the relative humidity reaches 100 percent—in other words, when the air is *saturated* with water vapor—the water must condense. Clouds of water and ice form, everywhere, and one can imagine the deluge. So long as volcanoes continue to spit out enormous volumes of water vapor, the air stays saturated, clouds pile up and burst, and water falls from the sky to form the oceans. The process here painted with the broadest of brush strokes certainly lasted much longer than forty days and forty nights. Even with 10 cm (4 in.) of rain per day, every day, everywhere, almost a hundred years would be needed to fill the oceans, provided there was no evaporation. And even if this deluge took a thousand or a million times longer, it would still be but a brief moment in the history of the Earth, a moment in which everything was forever changed. The geological record of sedimentary rocks, formed in the presence of liquid water, proves that oceans have existed for at least three billion years, maybe even four, and that their volume has hardly changed for the last two billion. Outgassing of the interior continues, but at a much reduced rate, most of the volatile material having already been lost to the surface and atmosphere. And the gases that emerge from volcanoes today result to a large extent from recycling between oceans and mantle.

What happened elsewhere? It appears that the Earth was just in the right place to keep its oceans.[11] Further away from the Sun and therefore receiving less sunlight than the Earth, Mars was colder. And with lower mass and density and a weaker magnetic field, the red planet had less internal ("areological" rather than geological) activity than the Earth; whatever gases were emitted by its now extinct volcanoes, they tended to condense frozen on the ground, so that having a much weaker greenhouse effect than the Earth, Mars remained cold. On Venus, closer to the Sun and very similar to Earth in size so that volcanic activity may have been intense, a thick atmosphere developed under much warmer conditions. Although the atmosphere must have contained water vapor initially, condensation of water was rare, and in the long run the water vapor was lost by the process of *photolysis*, solar radiation breaking apart the H_2O molecules. The hydrogen thus freed easily escaped the gravity of Venus. However, while on Earth most carbon dioxide has been locked up in the crust in the form of carbonates by way of chemical reactions that depend on the presence of liquid water, on Venus, without any ocean, this was not possible. On the contrary, the carbon dioxide accumulated as a gas in the atmosphere, leading to a runaway greenhouse effect, higher and higher temperatures, more and more CO_2 leaving the surface to join the atmosphere, reinforcing still more the greenhouse effect. For Venus, any chance of accumulating an ocean was lost.

A Living Solid Earth: Oceans, Sea-floor Spreading, Continental Drift, Earthquakes, and Volcanoes

We humans are landlubbers, living on continents and islands, land surfaces emerging more or less from the oceans. The 1,350 million cu. km of water *could* cover the *entire* globe with a single world ocean more than 2,500 meters deep; but in fact this mass of water is collected and partly separated in the seven seas, with average depth 3,730 meters. The biblical "All the rivers flow into the sea, and the sea is not filled; and the rivers continue to flow" in fact raises at least two questions. What we generally call the water cycle— evaporation, precipitation, runoff—does indeed explain that the rivers continue to flow, and that the sea does not overflow. But the rivers also carry with them the debris of the land: dissolved salts and minerals, 30 billion tons of mud and other suspended matter every year. How can it be that the *bottom* of the sea has not filled up? The facts are there: the sea floor is *not* a many-mile-thick layer of silt and other debris accumulated in the course of the billions of years since the origin of the Earth and its oceans. How can it be that the continents and mountains are still standing and not eroded and washed away by falling rain and running water? Not all these questions have been settled (are they ever, when scientific research keeps raising new ones?), but since the 1960s all these questions have appeared under a new light following the great revolution in the Earth sciences and the establishment of the discipline of *plate tectonics*.

In 1912 the German meteorologist Alfred Wegener (1880–1930) formulated the hypothesis of *continental drift*, according to which the continents move relative to one another, separating or colliding. Two hundred million years ago, Africa and South America were joined in one supercontinent, *Pangaea*—a long time ago, but still quite recent in terms of the age of the Earth. Using lasers on the ground to illuminate reflectors on board artificial satellites, space-age surveyors can make very precise determinations of the positions of those lasers, and the results confirm that Europe and North America are drifting apart 3 cm (a little over an inch) every year. The sea floor, always less than 200 million years old, is extruded from the mantle along a colossal underwater mountain range—the *mid-ocean ridge*—and spreads out in both directions until it finally plunges into the depths of the mantle of the Earth in the *subduction trenches* (fig. 2.1).

Maps usually color the continents differently from the oceans. This differentiation of the surface of the Earth—separation into oceanic and continental domains—remains valid even apart from the presence of the liquid water. The material of the continents was chemically separated early in Earth's history from the material of the sea floor. The not very dense continental crust is rich in silicates, but oceanic crust consists of basalt richer in minerals of iron and magnesium. The continental plates float above denser layers, like a sort of "scum," as distinguished French geochemist Claude Allègre put it.[12] Both the composition and the structure exhibit basic differences. It is striking that on modern maps showing the topography of the sea floor as well as that of the continents, the entire ocean realm is organized around an underwater mountain range much longer than continental cordilleras (over 60,000 km—i.e., one and a half times the circumference of the Earth). Starting from the Arctic Ocean

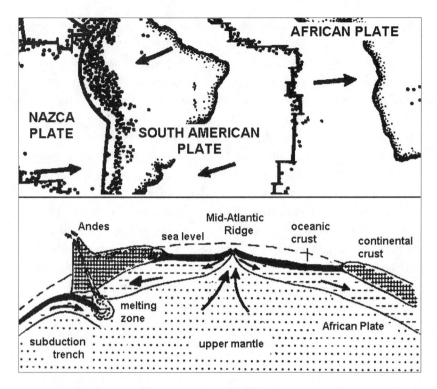

FIGURE 2.1 Sea-floor spreading and continental drift. At the top, a map of the tropical zone. From west to east: the eastern Pacific (Nazca plate), South America (Atacama subduction trench and the Andes), the Atlantic with the mid-Atlantic ridge separating the South American and African plates, and Africa. Dots indicate the epicenters of earthquakes. At the bottom, a vertical cross-section from the Nazca plate to Africa, with convective motions of the upper mantle indicated.

north of Spitsbergen, the mid-ocean ridge emerges from the ocean in Iceland. It then dives into the depths, and from north to south it divides the Atlantic into two roughly equal parts before turning east and passing south of Africa, the Indian Ocean, and Australia, finally extending north toward California and even Vancouver in British Columbia. A branch splits off in the Indian Ocean toward the Red Sea, while two others in the Pacific extend in the direction of Chile and the Galapagos Islands. The ridge, rising some 2,500 meters above the abyssal plain, with the surface of the sea about 2,500 meters above its summit, was hardly known before the 1950s. It differs from continental mountain ranges in its length, its composition, and its very peculiar structure: the crest of the ridge has a breadth of a few thousand kilometers, but it is split all along its length by a *rift*—a narrow valley (a few dozen kilometers), very deep in places, marked by numerous volcanoes and hot-water springs.

At present, and for the last several hundred million years at least, convection currents rise in the Earth's mantle (fig. 2.1) and culminate under the ridge, bringing up heat along with water trapped in the minerals of the magma, activating undersea volcanoes and hot springs. About 20 cu.km (several billion tons) of basaltic rock emerge around the rift and mid-ocean ridge each year. In about 200 million years, that amounts to enough to renew completely the fairly thin (5–10 km) ocean-floor crust. Where does all the rock go? Emerging as hot dilated lava at the rift, the rock cools, contracts, and becomes denser, spreading out on both sides of the mid-ocean ridge. How can we know this? Natural radioactive isotopes determine the age of the rock at various places across the sea floor, and fossil magnetic compasses of magnetite and other crystals record the direction of the changing magnetic field of the Earth at the time when the rock congealed. These measurements show that the sea floor is expanding by several centimeters (a few inches) every year, fast enough to open up the Atlantic Ocean in a hundred million years. This sea-floor spreading is possible because at depths ranging from 70 to 250 km beneath the ocean-floor crust, temperatures are around 1,000 K and the layers are partially molten and plastic, so that they do not resist lateral motions of blocks of crust floating above them. These layers are called the *asthenosphere*, the force-free or weak zone.

The rocks of the continental plates are less dense than the basaltic sea-floor crust, and so they float higher above the asthenosphere. Drifting like "rockbergs" carried along by the expansion of the floor of the Atlantic, North America and Europe move steadily farther away from each other, as do South America and Africa. As noted earlier, there was no Atlantic Ocean in a not so distant past: all the continents were clustered in a single land mass, Pangaea. This supercontinent was torn apart some 200 million years ago by the outbreak of the Mid-Atlantic Ridge, giving birth to the Atlantic. Other separations followed: the Indian subcontinent split off from Antarctica 135 million years ago, Australia was cast adrift some 70 million years later. But what was the situation earlier than 200 million years ago? Is it likely that Pangaea had existed in splendid isolation ever since the cooling of the Earth's crust billions of years before? There is some evidence that, on the contrary, other continents existed and drifted before then, meeting some 300 million years ago in a grand collision forming Pangaea. The sea floor is completely renewed every hundred million years or so, but some continental rocks are more than two billion years old. However, the continental blocks or plates can be broken, their pieces reassembled into new continents. As long as the Earth's interior remains hot, mantle convection will recycle the ocean-floor crust. The rift, where the "new" ocean-floor crust appears, changes over the course of time, with volcanic activity dying out in one place and erupting somewhere else, breaking up a continent and giving birth to a new ocean.

The continents drift as if they were made of rigid plates; and this also appears to be the case of much of the spreading sea floor. But where then does all the continually created sea floor go? After all, the Earth is round! A moving plate must ultimately run into another plate. Basically, there are two possible outcomes. Either one of the plates passes under the other and in so doing it returns to the mantle in a subduction trench; or else the crust is compressed and folded in the collision, with formation and uplift of a mountain range. At the bottom of the Pacific Ocean, the Nazca plate, having emerged from the East-Pacific spreading ridge, slides under the American plate in the Peru-Chile subduction trench along the west coast of South America, and in so doing it pushed up the Andes. Further north, the little Juan de Fuca plate slides under the coast from California to Alaska. To the west, the Pacific (sea-floor) plate slides under the continental shelf and pushes up the Aleutian and Kuril Islands and Japan. Behind these islands are only shallow seas, above the continental shelf: the Bering and Okhotsk seas, the Sea of Japan, and the China Sea. In some places, one oceanic plate slides under another: thus the Pacific plate plunges under the Philippines plate in the (11,000 m or 36,000 ft. deep) Mariannes trench.

Although the motions are slow, subduction entails friction between the plates, and also stress, strain, and heat. The waterlogged sediments accompany the ocean-floor crust in the subduction, and the presence of water facilitates the melting of rocks at depths of several dozens or hundreds of kilometers. Sometimes the stress on the crust is relieved by sudden slippage or breakage, an earthquake; the release of heat and gas keeps volcanoes active. This happens all around the "ring of fire" of the Pacific, from the intermittently active volcanoes (Mount Lassen, Mount Saint-Helens) of California, Oregon, and Washington state, north and west to Alaska and the Aleutian Islands, then to Japan by way of the Kamchatka Peninsula and the Kuril Islands, south to Indonesia and further on to New Zealand where the Pacific plate sinks into the mantle. A plate can also slide along another plate, along a fault, sometimes getting stuck. If the sticking point gives way all at once, the sudden shift of the plate produces an earthquake, as in California along the San Andreas fault. Along the west coast of South America, the stick and slip of the Nazca plate as it passes under the American plate triggers frequent earthquakes.

The continental plate of North America finds itself compressed in a sort of vise between the spreading of the Atlantic on the east and that of the eastern Pacific on the west. Perhaps this explains the uplift of the western third of the continent, from the Pacific coast to the Rocky Mountains. Does it also explain the extreme violence of the earthquake that took place in 1811, near the center of the continent and far from subduction trenches, mid-ocean ridges, or any major fault? This quake (at New Madrid, on the Mississippi, about halfway between Memphis and St. Louis) would cause a major catastrophe today, but at the time, population density was small and most construction light. As the African plate approaches Europe, progressively closing up the Mediterranean, the Alps have been lifted up, along with sedimentary rocks and fossil seashells. The effect of compression appears much more spectacularly in the uplift of the Himalayas and the Tibetan plateau, result of the collision of the Indian subcontinent with the Eurasian plate that began 55 million years ago, continuing to this day. Another spectacular example: the Caucasus Mountains of Armenia, Azerbaijan, and Georgia, resulting from the collision of the Arabian plate with Eurasia. In these areas, the high risk of collapse of buildings in stone (or in Soviet concrete, as in Armenia in December 1988)[13] makes quakes particularly murderous. Natural disasters are most disastrous when people are poorly prepared for them. It's been said that "one learns geology the day after an earthquake," but how many people remember the lesson?

Other dangerous encounters take place in the Caribbean, where the local Soufrière eruption buried a large part of the island of Montserrat in 1997, or around the Mediterranean, where a quake hit Assisi the same year, and where a city like Nice is at risk. Where volcanoes are involved, we are now able to avoid the enormous death tolls of past catastrophic eruptions, such as the eruption of Vesuvius (near Naples) that buried the cities of

Pompeii and Herculaneum in 79 A.D., or the eruption of La Montagne Pelée at 8:02 A.M. on May 8, 1902, which buried the city of St. Pierre de la Martinique (28,000 dead, only 2 survivors). It is now possible to avoid such death tolls, thanks to progress in the understanding of volcanoes and in techniques for closely monitoring their swelling, the seismic activity specifically associated with them, and the changes in outflow from hot springs. Indeed, a wide range of phenomena announce an impending eruption, even if these signs cannot tell geologists everything. It was clear that Mount Saint-Helens (Washington state) was going to explode in May 1980, and the area was evacuated, but there still were victims, in part because it was not foreseen that most of the explosive force would emerge to the side rather than upward. Today, the number of casualties resulting from volcanic eruptions can be kept small if we have the will, provided that the necessary observing stations and alert systems are established and maintained in the regions at risk, provided also that scientists and technicians can convince the public and public safety officials of the reality and seriousness of the risk. This has not always been the case, for example when a terrible mudslide wiped out Armero (Colombia) following the eruption of the volcano Nevada del Ruiz on November 13, 1985. Will Naples have to be evacuated one day? Neither Mount Vesuvius nor the nearby Phlygrean Fields are extinct. Mount Popocatapetl ("smoking mountain" in the Nahuatl language), only 60 km (36 mi.) southeast of Mexico City, occasionally makes people nervous, and with reason. An enormous catastrophe was avoided when the Philippine authorities evacuated the vicinity of Mount Pinatubo *before* it erupted in June 1991.[14] One of the factors contributing to this decision was the impressive documentary film put together by the Krafft couple. Volcano specialists from France, Katia and Maurice Krafft were among the 43 people who died when Mount Unzen (Japan) erupted on June 3, 1991. They knew the risks they were taking in staying close to the volcano.

On the seismic front, the search is on for ways to predict earthquakes. Along faults, regular gentle slippage of one plate relative to the adjoining one is probably the least dangerous course of events, and when monitoring shows that things are getting stuck, the red flags go up. Such monitoring can at least help risk assessment, even if it cannot reliably predict the day when a quake will take place. Some physicists, in particular the group in Greece known by their initials VAN, claim that certain electromagnetic wave signals can be detected just before a quake strikes. Whatever the merit of this particular method, it makes sense to look for advance signals of the sudden slippage or rupture under stress that constitutes the physical process of an earthquake. Nevertheless, the methods of prediction so far proposed are highly controversial: none is generally applicable, and even those for specific areas remain at best highly uncertain. Several million people were evacuated from the city of Haicheng (China) hours before a predicted earthquake struck in February 1975, but there was no usable prediction of the August 1976 T'ang-shan earthquake that took 700,000 lives. In any event, any false alarm, for example in Los Angeles, could have catastrophic consequences. And while people may be able to flee from threatened cities, their homes and workplaces cannot. Antiseismic construction codes and methods have proven their worth to some extent, but there still have been bad surprises, for example in Oakland, California (1989) and in Kobe, Japan (1995). It is hard to imagine any public or corporate decision to limit population growth and economic development in such high-risk regions as California, central Japan, or Istanbul, and so the highest priority must be given to making buildings more earthquake resistant and to teaching people how to react when a quake does strike.

Effective earthquake prediction remains elusive, but the international scientific community *has* succeeded in setting up an effective tsunami alert system, at least around the Pacific Ocean. The word *tsunami* comes from the Japanese, and it stands for big waves that wash over coastal areas. They are caused by underwater earthquakes or mudslides, and not by tides, although they are sometimes called "tidal waves." The wave is triggered by the sudden shock, shift, or vibration of the ocean floor, and it then propagates in all directions at a speed in the range 700–1,000 kilometers per hour (400–700 mph). In the open ocean, the wave, only a few centimeters (1–2 in.) high with wave crests about 100 km (60 mi.) apart, cannot be seen. The wave continues its travel for thousands of kilometers, but with decreasing ocean depth close to the coast, wave height can grow to tens of meters. Triggered by an earthquake off the coast of Chile on May 22, 1960, causing thousands of casualties in Chile and Peru, a tsunami reached Hawaii fifteen hours later, wiping several coastal villages off the map and partly destroying the city of Hilo; six hours later still, the same tsunami reached Japan, causing hundreds of casualties. On August 27, 1883, the volcano Krakatoa (Indonesia) exploded with a force of 100 megatons of TNT (as big as or bigger than any H-bomb ever tested): this triggered a local tsunami with waves 40 meters (130 ft.) high, drowning 36,000 persons.[15] For local tsunamis, there is not enough time for warning, but an alert can be given wherever the tsunami has to travel far from the initial shock. Seismological measurement stations the world over immediately inform the International Tsunami Warning Center outside Honolulu of the location and strength of any seismic tremor likely to trigger a tsunami. The center computes the tsunami's travel times to the different Pacific coasts and issues the warning. Given the size of the Pacific, those travel times can be several hours long, and effective precautions can be taken. Elsewhere—for example, in the Caribbean or the Mediterranean—advance warning time is much shorter. In 1908 a Mediterranean tsunami drowned thousands in Messina (Sicily) and Reggio di Calabria. If you are near the seashore and have the impression that the sea is suddenly emptying out and retreating far far away, you had best head for the hills, or at least higher country, as fast as you can, because the sea can come back in an enormous wave that can wash everything away over hundreds of yards.

For the most part, earthquakes and volcanic eruptions occur either along the mid-ocean ridge, or along the lines where plates collide or slip past one another. These tremors and exhalations are the superficial signs of the convective currents that stir up the mantle, with magma rising at the mid-ocean ridge and ocean-floor crust descending in the subduction trenches. Other volcanoes and other earthquake sites are to be found at what are called "hot spots," isolated points where magma comes up through the crust. One of these is the Kilauea caldera near the southeast coast of the big island of Hawaii. Tourists can leave an orchid there, at the edge of the vast Halema'uma'u crater, in homage to the goddess Pele. And if the continuing exhalations of sulfur dioxide (SO_2) and of hydrochloric acid (HCl) do not bother them too much, they can descend into the crater, taking care not to stray from the marked trail because in other places, the recently solidified glassy crust can break under their weight. The crater was full of molten lava when Mark Twain visited it in June 1866, and it could fill again in a coming year. Lava is emerging today from the Pu'u O'o crater, a bit closer to the sea. The lava flow cut the road and buried the Waha'ula temple built some eight centuries ago, and it plunges into the sea, generating an enormous cloud of steam laden with acid vapors and glassy dust. Off the coast, 32 km (20 mi.) to the southeast, a new island already given the name Loihi is being born: its peak is still a thousand meters below sea level, but it will emerge in 50,000 or 100,000 years. Returning to the center of the big island, one can climb the slopes of Mauna Loa, a volcano that since 1960 has been relatively quiescent. But Mauna Loa is not extinct, and in 1942 a lava flow threatened the city of Hilo. Further to the north, astronomers often visit Mauna Kea ("white mountain" in Hawaiian). With an altitude of 4,200 meters (14,000 ft.), the peak is often snow-covered. Well-equipped observatories have been established there to take advantage of the clean thin air, funded by many different countries (in particular the Canada-France-Hawaii 360-cm or 142-inch telescope). Obviously, the astronomers hope that the volcano is extinct. Both Mauna Loa and Mauna Kea rise more than 8,000 meters above the nearby ocean bottom. Mau'i with Haleakala, Oahu with Diamond Head, indeed all the islands of the Hawaiian chain out to Midway 2,000 km (1,200 mi.) west of Honolulu, mark both the volcanic activity of the past and the spreading of the Pacific ocean floor. What seems to have happened is that a fixed spot has from time to time heated up and erupted, forming one or more volcanoes, building up a new island. And then the motion of the sea floor has slowly but inexorably carried the island off to the west, removing the volcano from the "hot spot" and thus extinguishing it. With the next spurt of hot-spot activity, a new volcano is born, a new island built up. Elsewhere, in the Pacific, in the Indian Ocean, and in the Atlantic from São Tomé to Mount Cameroon, strings of volcanic islands mark the spurts of activity of hot spots under a moving ocean floor. Other hot spots exist under the continents—for example, the one that heats up the geysers and hot springs of Yellowstone.

From the mouths of a thousand volcanoes, the Earth spews out lava, steam, carbon dioxide, sulfur dioxide, hydrogen sulfide (H_2S), hydrochloric acid, and more. In a distant past, fits of volcanic frenzy buried vast areas with lava, in the Deccan Traps of India, and the Columbia River basalt of the northwest United States. On the continents, such solidified lava discharges last long, but on the ocean floor they are recycled in the mantle by way of the subduction trenches. The lava emerging today in the rift or in newly opened craters on Hawaii or Montserrat may contain minerals already present in the sea floor of ancient oceans. The steam spouted by volcanoes today comes for the most part from recycled water that was included in the sediments descending in the subduction trenches, water that was for a time lost to the oceans and atmosphere. And so, even though water vapor has emerged from volcanoes for billions of years, the seas are not full.

From the ocean bottom through the sea floor, water is drawn into a still deeper water cycle, reappearing in more or less vigorous volcanic emanations. But apart from geysers, this subsurface cycle can only supply the sources of rivers by way of the atmospheric branch of the water cycle. Some of the steam emerging from Pu'u O'o on Hawaii may conceivably fall as rain on the Kahlua coffee plantations of the western slope of the island the same evening, rejoining the waters of the Pacific by week's end. Each raindrop of the showers that water the apple orchards of Normandy includes a few H_2O molecules coming from the Soufrière, Etna, and Kilauea volcanoes, among billions of billions of water molecules evaporated from the surface of the Atlantic Ocean. And those billions of billions may include a few molecules recently emerged from the mid-Atlantic ridge. Some of that water will be incorporated in the apple that you eat on a visit to Paris. The rest will drain into the English Channel, the Atlantic Ocean, or the North Sea, unless it is reevaporated. What does evaporate can condense to form a cloud, and a few days later may water the Tokay (Tokaj) vineyards in Hungary and run off in the Danube River to the Black Sea. Whenever you eat an apple or drink a glass of wine, you are absorbing water that has cycled through the atmosphere thousands of times since you were born. But you are also absorbing some water molecules that have only been out in the open air for a few days or weeks, after tens or hundreds of millions of years beneath the Earth's crust.

CHAPTER 3

ORIGIN AND EVOLUTION OF LIFE

An Extraterrestrial Origin?

At the turn of the century (around 1900), the Swedish chemist Svante Arrhenius[1] hypothesized that life spreads throughout the universe in the form of *spores*, drifting from one planetary system to another, driven by the radiation pressure forces of starlight. On Earth, spores are extremely hardy forms taken by bacteria, fungi, ferns, and mosses, allowing them to survive under conditions completely hostile to life as they await better times and climes to revive and multiply. Arrhenius imagined that sporelike forms of life could survive for very long periods under the harsh conditions of interstellar space, providing the seeds of life once they fall on a suitable planet. This so-called *panspermia* hypothesis is still entertained by some scientists. One variant would have the spores arrive as passengers of comets rather than as individual drifters. Others would have them arrive with spaceships directed by extraterrestrial intelligent beings: either as deliberately planted "seeds" of life, or as accidentally leftover crumbs from a picnic. But even if such a hypothesis of terrestrial life arriving from somewhere else is credible, all it does is to change the locale of life's beginnings. According to this hypothesis, the first chickens emerged from eggs from outer space (or rather, the first terrestrial microbes arose from spores that arrived on our planet from outside the solar system); but is this a satisfactory solution of the chicken-and-egg problem? In our universe, hydrogen and helium were practically the only elements in existence when the first stars began to shine; how then did life, made mostly of heavier elements, begin?

One way of avoiding the question, not of answering it, is to deny that there ever was a beginning of either life or the universe. In the 1950s, the English cosmologists Hermann Bondi, Thomas Gold, and Fred Hoyle argued for a steady-state universe with continuous creation rather than a Big Bang, and some cosmologists still argue that this theory can be revived and made compatible with the observations. But even if the seeds of life have always existed, somewhere in the universe, we still need to explain how on Earth life sprouted from such seeds, and how it spread and diversified over the past four billion years.

MATTER LIVING AND INERT

What then is life? In the vast diversity of living species, we can pick out some characteristics shared by all. From the smallest bacterium to the tallest tree, from the flea to the whale, all living creatures are built around water. Life is to be found at all altitudes on the surface of the Earth or in the depths of the oceans, at temperatures ranging from −50 to +50°C (say from −60 to +125°F). Some worms are found in glacial ice, and they must feed on bacteria themselves adapted to the cold. That such psychrophilic (cold-loving) bacteria exist constitutes a risk in cold storage of food. At the other extreme, thermophilic (warmth-loving) bacteria do well at high temperatures, with hyperthermophilic bacteria surviving and indeed thriving at temperatures above 100°C (212°F, the boiling point of water at atmospheric pressure). They are found in the boiling water emerging from hot springs, as in Yellowstone National Park, and indeed at much higher temperatures deeper down where the water can remain liquid because of the higher pressure.[2]

Although water is the medium of living matter, it does not in itself define the structure of life. In all living creatures large or small, the fundamental attributes of life, namely metabolism and reproduction, depend on and reside in the complex molecular structures that make possible the storage and transmission of genetic information and of energy. Although such structures do contain hydrogen and oxygen, constituents of the water molecule H_2O, more importantly they contain carbon and nitrogen. Putting together atoms C, N, O, and H in different ways and proportions, nature comes up with a wide variety of robust molecular structures. These start with quite simple molecules—*monomers* such as ethylene (CH_2)—assembled into more and more elaborate constructions known as *polymers*. *Polyethylene* is made of hundreds of linked ethylene monomers, while other polymers are assembled from different monomers. Thus the amino acids, essential components of proteins, consist of an NH_2 radical (i.e., an ammonia molecule NH_3 having lost a hydrogen atom), linked up with other fairly simple molecules made up of carbon, oxygen, and hydrogen (C, O, H). Examples include ethylene (CH_2) and carboxylic acid (COOH). Three of the amino acids also contain sulfur (S). Each amino acid molecule represents a little structure made up of five to twelve monomers, each in turn containing one to four elements, usually C, N, O, and H. Indeed, these four elements constitute 99 percent of the mass of living beings. Adding a dash of sulfur and phosphorus makes it possible to complete what is sometimes called the biochemical "alphabet"—29 molecules from which the even more complex structures of life can be assembled. In addition to the proteins, these include DNA and RNA (deoxyribonucleic and ribonucleic acids), agents essential for transmission of genetic information, as well as the ATP molecule (adenosine-triphosphate), needed to convey energy within living cells. And we all know, even if we don't empty the shelves of "health food" stores, that small amounts of other elements remain essential: the sodium and chlorine found in salt, as well as calcium, magnesium, iron, manganese, iodine, zinc, and so on. Still, the basic structure of life is built around C, N, O, and H—even if only calcium-rich skeletons and shells are all that are left in the long run, after death and decay.

Some 92 elements are found in nature, but it takes less than two dozen to put together the multitudinous molecules of life. In the nineteenth century, chemists would invoke a "life force," making a fundamental distinction between "organic" chemistry (i.e., the chemistry of carbon compounds) and the "inorganic" chemistry of all the rest. Today, chemists synthesize "organic" molecules in the laboratory and in pharmaceutical factories, without life playing any role. Fairly elaborate molecular edifices also take shape far from the planets. Radio astronomers have identified over a hundred organic molecules including some amino acids, some of them containing as many as fifteen atoms, in cold and relatively dense interstellar clouds. Mother Nature knows how to make amino acids in the desolation of interstellar space, humans have learned how to do it in the laboratory. In 1953 the chemist Stanley L. Miller, working at the University of Chicago with Harold C. Urey,[3] synthesized several different organic molecules, including amino acids. They did this with electrical discharges in a mixture of water, hydrogen, methane, and ammonia that Urey considered to be representative of the Earth's primordial atmosphere, similar to the atmospheres of Jupiter and Saturn. But did such an atmosphere ever exist on Earth, and if it did, could it have lasted long enough for life to get its start? Other investigators have shown that amino acids can also be produced starting with other mixtures—for example, with water, nitrogen, carbon dioxide, carbon monoxide, and ammonia—corresponding to what may well have been the *secondary* atmosphere of the Earth. Volcanic gases also contain substantial amounts of hydrogen sulfide (H_2S), from which the amino acids containing sulfur can be formed. Nobody can pretend today to know exactly how things proceeded, but it seems reasonable to believe that the inherent self-organizing properties of matter necessarily and inevitably led to the accumulation of a stock of "prebiotic" matter (i.e., of the complex organic molecules needed for the formation of the first living beings) at an early stage in the long history of our planet.

How can this self-organizing process be described? In my personal view, neither the term of "a biochemical alphabet" nor that of "the bricks of living matter" fits the bill. Of course, with an alphabet of 26 letters, 10 figures, and a few other signs, I had all I needed to write this book, not to mention an encyclopedia; but that doesn't explain why certain arrangements make sense (form words), while others do not. Nor is it because I have specified that DNA stands for "deoxyribonucleic acid" that I have explained what it is. Using the alphabetical symbols for the elements, chemists write H_2O for the water molecule, and that tells you something if you remember what H and O stand for; but to know more, you need to know the rules governing the orbits of the electrons around the nuclei of these atoms, and the way in which they can form links. As for bricks, architects and masons can certainly use them to create remarkable edifices. Nevertheless, the properties of a brick hardly determine the elegant features of a seventeenth-century Amsterdam burgher's house, or the sadder traits of Parisian low-rent housing of the 1930s, or of a New York housing project of the 1950s. Moreover, to hold a brick building together, mortar is needed. It is true that the ancient Romans and Incas knew how to choose and cut stones and build long-lived structures without using mortar, but although the "building blocks of life" are more varied than simple bricks, they are nonetheless standard elements and not custom-cut. Rather than an alphabet, rather than a pile of bricks, I would suggest that the best metaphor is a sort of Lego set, where the different pieces, not identical as bricks, fit one into the other in different ways. Imagine atoms to be Lego pieces of different sizes, with fewer or more nubs and cavities where they can be fitted together. Assembling thus a few of these pieces to form relatively compact structures of limited size gives simple sturdy molecules—the monomers. These can in turn be put together into larger, more elaborate, often three-dimensional structures, no doubt less robust but still not fragile by any means. Such structures are certainly not bricks; often they have features more like the holdfasts or the mortise and tenon joints of cabinetmakers. No plan is needed for assembly, and most of the structures are incomplete. However, the residual electric charge on one side of an incomplete structure necessarily attracts other structures with the complementary opposite electric charge. In a sense, most of these molecules and radicals have "hooks" and/or "eyes," and they tend to hook up in chains, rings, in chains of rings, and so on and so on. In this way, the basic electrical forces and properties of matter lead naturally (how else?) to the formation of giant molecules—*macromolecules*. Are you shocked by such a description of the formation of the basic constituents of life in terms of a sort of spontaneous Erector set? Remember, this deals with the assembly of the material of life, not of its destiny. Whether you use the Latin or Arabic or Hebrew alphabet to write, or for that matter Chinese ideograms, you had to start (at least before computer keyboards!) by learning to draw and assemble diversely oriented lines and curves; that in itself does not determine whether you write a love letter, a poem, or a technical manual. However, from the microbes that Leeuwenhoek[4] discovered in a drop of water, to the diversity to be found in a tidal bay or a tropical rain forest, life's saga seems to have written itself, as though all of Shakespeare's plays emerged from a typewriter without even a monkey to type them.[5]

How was the first DNA molecule assembled, starting from "prebiotic" organic molecules such as amino acids and sugars? DNA, a truly gigantic macromolecule, is built up from *tens of millions of nucleotides*, each one being a molecule composed of the sugar deoxyribose (made of carbon, hydrogen, and oxygen), a nitrogen base, and a phosphate. For DNA, there are four types of bases (for example, adenine), made of five nitrogen and five hydrogen atoms. The DNA structure, a double helix[6] composed of two linked chains, explains the reproduction of the genetic code: the linked chains can separate, and each then acts as a template from which the other chain is regenerated. Biologists still don't know when and how the first DNA molecule was assembled. As for the different kinds of RNA, which are needed for the production of proteins and for transmission of energy as well as genetic information, the difficulties are similar, but the molecule is simpler—a single chain rather than a double helix—and much smaller (1,000 to 50,000 nucleotides).

Three-dimensional structures, such as those of these macromolecules, can exist in two forms, one being a mirror image of the other. Life on Earth is based on one rather than both of the forms of each such molecule. Nearly all the amino acids exist in the "left-handed" form, so called because when a polarized light beam passes through the acid, the plane of polarization rotates leftward; nearly all natural sugars take the right-handed form. This dominance of one of the forms must have appeared quite early in the history of life, although nothing in the intrinsic properties of these molecules favors one form over the other. "Mirror-image life," with its 3-D molecules in the opposite form, can be imagined. However, it is in the nature of life to reproduce itself, always following the genetic code or instruction set in the DNA structure. The proteins are produced by the template of the RNA structure, with the "key" molecules fitting into the corresponding "lock" molecules. Unless *all* the molecules are mirror images, the mirror structure will not work. Sheep made of "mirror" proteins can only digest "mirror" grass, and we would find them quite indigestible; and man-eating "mirror" tigers could only digest "mirror" men (and women). Even if you can get your left shoe on your right foot, it won't be comfortable. In the laboratory, amino acids can be produced both in right-handed or left-handed forms, as Louis Pasteur (1822–1895) found when he first synthesized tartaric acid and its mirror in the middle of the nineteenth century. Today, when chemists in the pharmaceutical industry synthesize new molecules, they must check whether the form produced is more useful (or less deleterious) than its opposite.

Maintenance of biological structures requires energy. Today, and for much of the history of life on Earth, nearly all the energy comes from the Sun. By the process known as *photosynthesis*, green plants harness the energy of sunlight, producing glucose and other organic molecules from water and carbon dioxide. They're known as *autotrophs* because they produce their own organic matter using the abundant inorganic molecules H_2O and CO_2 as raw material. Glucose has a fairly simple chemical formula ($C_6H_{12}O_6$), showing the same ratio of hydrogen to oxygen atoms—2 to 1—as in water. Once organic molecules such as glucose have been fabricated by photosynthesis, the process of *respiration*, using atmospheric oxygen, releases some of the converted solar energy by oxidizing some of the organic matter, in effect *burning* it, thus reconverting it to CO_2 and H_2O. Living beings called *heterotrophs* feed either directly (for example, in the case of herbivorous animals) or indirectly (carnivores) on the organic matter of green plants and other autotrophs. This provides today's heterotrophs the organic matter of which they are made, with energy of solar origin stored in the organic molecular structures. Animals, birds and bees, worms and weasels, frogs and fish, all get their energy from respiration, burning carbohydrates using the oxygen in the air or dissolved in water.

Such metabolism depends on the availability of oxygen, today about 21 percent of the atmosphere. It could not have worked for the first living things, because the fraction of oxygen in the Earth's atmosphere was then very small, rising above 1 percent only fairly recently (2.2 billion years ago?). Life without oxygen? In fact, many bacteria are *anaerobic*, thriving in the absence of oxygen, for example in our intestines or in those of cows, or in swamps. Indeed for some of them, oxygen is a poison. Some organisms more complex than bacteria get their energy from the fermentation process, with no need for oxygen even though it does not indispose them. An example: the microscopic fungi (mushrooms) of yeast used since millennia to prepare leavened bread. However, fermentation operates on ready raw material, already formed organic matter such as glucose. When the glucose ferments, alcohol (a somewhat simpler organic molecule) is produced, and carbon dioxide and some energy are released. The released CO_2 raises the dough when preparing biologically leavened bread, the heat of the oven drives off the alcohol; when brewing beer, both are retained.

Where and When Did Life Appear?

To find the first living things, one must look for much simpler organisms than yeast: such cells are already quite complex, having nuclei, so that they belong to the class of *eukaryotes*. Nevertheless, yeast is an everyday reminder that life forms can exist without oxygen and without photosynthesis, provided that they can extract energy from a ready stockpile of organic molecules. Photosynthesis, with its complex chemical processes, almost certainly represents a step in evolution that took place definitely after the beginning of life on Earth. The very first living things must have got their energy from prebiotic molecules accumulated in the early ages of the Earth, or even before, these molecules storing energy of solar, geothermal, or perhaps even extra-solar-system origin. There still exist today extremely simple micro-organisms, constituted by a single cell without any nucleus. Called *prokaryotes*, some of these require neither oxygen nor photosynthesis. Fossil prokaryote bacteria (about 0.01 millimeters or 0.4 mils or thousandths of an inch in size) have been found in deposits two billion years old near Lake Superior, and some of them are quite similar to species living today. Some still older microfossils have been found in sediments over 3.3 billion years old in Australia and in South Africa, although the metabolism of these bacteria is not clear. According to some researchers, the carbon isotope ratios (C-12/C-13) of the very oldest sedimentary rocks (3.8 billion years, in Greenland) betray biological activity.

Today, along some seashores warmed by the Sun and regularly washed by the tides—for example, the beaches of the Persian Gulf, of the isle of Andros in the Bahamas, of Shark Bay in northern Australia—blue-green algae (cyanobacteria) grow, forming regular structures with the mud that are shaped by the strong tidal currents that keep out the animals for which the algae are a tasty treat. These matlike structures are called *stromatolites*, from the Greek words *ströma* for carpet and *lithos* for stone. Fossil stromatolites are frequently found among the oldest rocks of the planet, of the Precambrian, more than 3.5 billion years old, in Canada, South Africa, and Australia. They are less common in more recent rocks, probably because greedily grazing marine invertebrates appeared in the Paleozoic some 570 million years ago. Nevertheless, these blue-green algae have managed to survive in areas where strong currents sweep away the animals that would feed on them.

However, photosynthesis by algae is a step beyond the beginnings of life. Did life start off with archaic anaerobic bacteria in a "little warm pond," as Darwin put it? Or in a warm ocean, saturated with prebiotic molecules, a sunny primeval soup? The idea has been around for many years, although the phenomenon itself has never been demonstrated. Indeed,

Pasteur's germ theory came with the demonstration that "spontaneous generation" of life does *not* occur today. Stanley Miller's early results are often cited: starting with a reducing atmosphere and activating it with energy from ultraviolet radiation or electric discharges, prebiotic molecules are produced. Sometimes the argument is turned around: since we know that life exists, the early atmosphere of the Earth must have been as Miller and Urey supposed. However, some dissenters—for example, the French biochemist Antoine Danchin[7]—argue that such a prebiotic soup would have been a "poisonous potion," and that life must have got its start in clay, on dust grains, or on the surfaces of rocks, perhaps before the Earth was formed. In any case, there exist even today very simple bacteria (archeobacteria) that need neither oxygen nor energy from the Sun to stay alive. Some get their energy from fermentation, others have a sulfur-based metabolism. Perhaps life started off in hot springs, deep down. But if so, how did it emerge to spread and thrive in the cooler oceans and on the continents? Recent discoveries made on the sea floor suggest still another scenario of life independent of the Sun.

LIFE WITHOUT THE SUN AT THE BOTTOM OF THE SEA

Today, for the most part, marine life thrives near the sea's surface, where the Sun's rays can penetrate, especially in those areas where the waters are enriched in nutrients stirred up from the sea floor by upwelling, or brought from the land by river flow. By photosynthesis, phytoplankton (algae) use the energy of sunlight to transform carbon dioxide and water into organic matter. Starting with these maritime "meadows," the food chain goes through the zooplankton (microscopic sea animals) that graze on the phytoplankton, through the little fish and big fish as well as dolphins and whales that feed on them, out into the open air to seagulls, cormorants, and other birds, and of course to humans, sometimes by way of fishmeal, fowl, and beef. On the sea bottom, mollusks and crustaceans feed on the debris of near-surface life, crumbs of the solar feast. In general, relatively little life is found on the pitch-dark deep-sea floor, for not much is left to eat at those depths, except for the rare carcasses of dead whales.

In 1977, exploring the undersea rift near the Galapagos Islands with the research submarine *Alvin*, oceanographers were absolutely astounded to discover teeming living communities at depths of 2,500 meters (8,200 ft., 1,367 fm or fathoms). Since then, many other communities of this type have been studied along the branches of the mid-ocean ridges of the Pacific as well as along the mid-Atlantic ridge. All along the ridge (fig. 2.1), the volcanic rift valley is dotted with springs from which water emerges, warm (2 to 30°C, i.e., 36 to 86°F) or hot (200 to 350°C, 392 to 662°F), enriched in sulfur and other compounds. These "hydrothermal vents" are surrounded by remarkable creatures—species of clams, crabs, worms, and mussels— never seen before. And what an abundance of life! Whereas there is seldom more than 500 grams per square meter of biomass in fertile and productive estuarine ecosystems, virtually never more than 10 grams per square meter on the deep-sea floor, these hydrothermal vent ecosystems sometimes have as much as 10,000 grams of biomass per square meter. All of this is, however, concentrated in a small area (not many square meters) around the vent, much smaller than the areas occupied by coastal or estuarine biomes. But what do these worms, crabs, and giant clams live on?

The energy comes not from the Sun but from the Earth's interior. Very hot water (300 to 400°C, i.e., 500 to 700°F, liquid because of the high pressures) flows out of deep pockets of magma (molten rock) at 1,200°C; forced through the Earth's crust and reacting with the rock, it emerges in the cold ocean water (2°C, 36°F) loaded with dissolved and chemically reduced minerals: sulfides, metallic ions, manganese, iron, and so on. As this supersaturated water cools on contact with the cold ocean water, some of the minerals, in particular the sulfides, cannot stay dissolved. The precipitating pyrites (iron disulfide) make the water black, appearing as spectacular "black smokers," with very hot water sometimes streaming out at several cubic meters or tons per second, depositing sulfides which accumulate to form chimneys tens of meters (yards) high. Where water is not quite so hot, "white smokers" appear; and even when the water appears clear, it is heavily loaded with dissolved minerals. This mineral-rich warm or hot water supports very special forms of autotrophic life, bacteria getting their energy from the dissolved carbon dioxide and minerals—"chemosynthesis" rather than photosynthesis. In fact, the energy emerges from the Earth's interior not only as heat but also by way of the chemically reduced state of the minerals dissolved in the hot water. The bacteria extract potential chemical energy by oxidizing these molecules and ions in different ways, and build up their own organic matter. This chemosynthesis is the start of the food chain that nourishes the entire hydrothermal vent biological community. Some of the bacteria exist in especially close symbiosis with animals of the vent community, exchanging oxygen, carbon dioxide, and sulfides with them, to the extent that some of the animals don't even have a mouth or other organ for ingesting solid food. Particular species of blue-green algae or cyanobacteria cover the sea floor around the vent, directly or indirectly providing nourishment for the grazing animals or predators (crabs, snails, fish, shrimp, etc.). One of the shrimp species has no eyes but rather organs sensitive to the infrared radiation of the smokers. And while sulfides are poison for many "normal" animals, they are well tolerated by the species gathered around the hydrothermal vents. What is the origin of these amazing communities? And how can they survive, considering that each individual hydrothermal vent seems to stay active for a few decades only, and each vent community is separated from others. Adapted as it is to the very special conditions surrounding the smokers, how can such an ecosystem survive and spread through the cold ocean over the barren deep-sea floor?

The discovery of these communities based on chemosynthesis, with a food chain going from the chemotrophic bacteria all the way up to a few specialized predatory fish, has suggested an alternative scenario for the beginnings of life—not in a warm sunlit pond, but in the outpouring of heat from the Earth's interior into the cold dark depths of the sea. The inferno of each black smoker nourishes and shelters an oasis of life. Indeed oceanographers have given the name "Genesis" to one of the hydrothermal vent communities of the East Pacific rise. Did life begin around a black smoker, only later colonizing the rest of the planet and eventually adapting and taking advantage of the availability of abundant solar energy? Or, on the contrary, do the teeming vent communities result from the adaptation of some hardy individuals that strayed away from the sunny surface layers of the sea? And they must have been hardy indeed to survive in the boiling toxic sulfide soup! No one knows. And even if life began at the bottom of the sea, the living creatures found there today no doubt include some that evolved in part under the Sun as well as others that never left the abyss. Furthermore, in some places on our planet, the two environments exist side by side: the mid-ocean ridge with its volcanoes emerges at the surface in Iceland and in the Horn of Africa. Somewhere on Earth, four billion years ago, the volcanic rift may well have been near the sea's surface.

THE INVENTION OF PHOTOSYNTHESIS AND THE CONQUEST OF DRY LAND

What else has happened in the past three billion years? With a single "invention"—photosynthesis—Mother Nature gave life the tool with which to spread, to diversify and to endure, while radically transforming the physical and chemical environment. The first micro-organisms, with ample prebiotic matter available in their environment, could prosper and multiply thanks to the genetic code of DNA, the key to reproduction. Ultimately, however, they would have exhausted this stock of raw material, a nonrenewable resource, unless they happened to be next to some sort of natural "recycling plant." Hot black smokers act in such a way, producing chemically reduced minerals, but they could never have provided enough energy to populate the planet. The evolution of photosynthesis made it possible to tap the vast reserves of carbon locked up in the carbon dioxide of the atmosphere. With photosynthesis, a complex chain of chemical reactions uses the energy of sunlight to break up the H_2O and CO_2 molecules, forming organic molecules (carbohydrates) and molecular oxygen (O_2). Although the solar energy resource will eventually run out, it still has several billion years to go. Today, after three billion years of photosynthesis, our planet's stock of inorganic carbon (CO_2) has been diminished. Even after two centuries of fossil fuel burning, the amount of CO_2 present in the atmosphere is quite small; much more is dissolved in the oceans. Small as it is, the stock of CO_2 in air sustains life on Earth: the living biosphere continually transforms CO_2 into organic matter by way of photosynthesis, using solar energy, releasing this energy by oxidation, and returning most of the CO_2 to the atmosphere by way of respiration. Most of the organic carbon on Earth results from the accumulation, over hundreds of millions of years, of a small excess of photosynthesized organic carbon, buried before it could be oxidized. This excess includes the fossil fuels—coal, oil, natural gas—which we humans are rushing to burn up in a few centuries, especially since 1950, sending the CO_2 back into the atmosphere.

Many people use the term *pollution* to refer to these emissions of carbon dioxide resulting from fossil fuel burning. Some even call them "toxic." It is true that any gas in excess can be toxic, even oxygen. However, although carbon dioxide is suffocating if it completely displaces air (as happened when a huge CO_2 bubble escaped from Lake Nyos in Cameroon on August 21, 1986), it has no untoward effects on humans or animals at anything close to present or near-future atmospheric concentrations (374 ppm or parts per million, i.e., 0.037% in the year 2000, rising to perhaps four times more in the next two hundred years). Substantially higher CO_2 levels are often reached in meeting rooms or even worse in airline cabins, but the discomfort of such environments is only partly due to the high CO_2. However, carbon monoxide (CO), produced by incomplete combustion of carbon, is truly toxic even at very low concentrations. In the atmosphere, on average, CO is present at 0.12 ppm; but the high CO levels often observed in cities (that were also quite common around the campfires of our cave-dwelling ancestors) constitute real pollution and a danger to health.

Carbon dioxide must have been a significant constituent of the Earth's atmosphere, present at levels above 10 ppm, for nearly all the history of our planet. Many geochemists are of the opinion that over most of the last 700 million years, the CO_2 level was as much as a hundred times higher than it is today, and that it was still higher before then. Some scientists believe that concentrations higher than 100 ppm are necessary for the continuation of photosynthesis and the survival of the biosphere in anything like its present state. It is known that CO_2 concentrations fell below 200 ppm during the coldest phases of the ice ages, the last being only 18,000 years ago. I personally find that because of its essential role in the maintenance of life, CO_2 does not deserve to be called a pollutant, but this is not to discount the seriousness of the problem of the climate changes that will result if atmospheric CO_2 continues to increase over the coming century.

Looking at the truly long term gives a different perspective on pollution. When life began, there was practically no oxygen in the atmosphere. For the first micro-organisms, oxygen was a highly toxic gas, and even today many anaerobic bacteria cannot tolerate it and only survive in oxygen-free environments. From the point of view of most of the ancient biosphere, the first organisms practicing photosynthesis—no doubt the ancestors of today's blue-green algae—were in fact polluting the environment with their poisonous oxygen emissions. Life found ways to adapt. Variation and natural selection have led to the present situation. Whereas some species called facultative anaerobes only tolerate oxygen, many others actively exploit it with respiration (oxidation) to release energy. Still other species can survive only in anoxic environments, where they can play an essential role, an example being the microbes that help to digest cellulose in the innards of cattle and termites. Is it shocking to read this description of oxygen as a pollutant? Today, we consider oxygen to be essential for life, with forests often described as the "lungs" of the Earth. Still, excess of oxygen is dangerous for the eyes of newborn infants as well as at the level of the living cell. In the atmosphere as a whole, oxygen is diluted by nitrogen. There is little chance of it reaching levels much above the present 21 percent because such an increase would trigger more frequent and more extensive forest fires, bringing the oxygen concentration back down.

The enrichment of air with oxygen seems to have accelerated about 2.5 billion years ago, with most of the "red beds" of iron oxides having been laid down since then. This must have gone along with the proliferation of the first photosynthetic organisms. In photosynthesis, the energy of certain wavelengths of sunlight serves to break apart the H_2O molecules, and molecules of glucose are assembled by linking the hydrogen thus freed to atoms of carbon and oxygen from atmospheric CO_2; at the same time, oxygen (O_2) molecules are freed. Of course, one might imagine other processes not involving photosynthesis breaking apart water molecules, leaving oxygen behind while the freed hydrogen molecules or atoms escape from the Earth's atmosphere because of their low mass. However, such processes are relatively ineffective, and in any case the oxidation of exposed rocks must have kept the oxygen level very low.

With photosynthesis, it's a whole new ball game. Existing at first only in the seas, photosynthetic algae must have released oxygen, while part of the organic material ended up sinking to the bottom where it was effectively isolated from the atmosphere for a long time. Without such isolation, organic matter in contact with the air's oxygen would be oxidized, and this reaction would keep the oxygen level down. As long as the atmosphere contained little oxygen (O_2), practically no ozone (O_3) could be formed, and as a result the atmosphere could not have blocked much of the Sun's ultraviolet radiation. The photons of such radiation individually transport enough energy to disorganize the macromolecules of living matter, in particular DNA. The conclusion: early life was restricted to a medium, such as water, shielding it from the ultraviolet but allowing penetration of the wavelengths used in photosynthesis. With their oxygen emissions, the ancient algae must have progressively polluted the atmosphere, making it more and more noxious for most of the microbes of the time. This led to the development of oxygen-resistance and of mechanisms taking advantage of the oxygen, and may have had something to do with the appearance of cells with nuclei (eucaryotes) and then of the first multicellular organisms. By 1.3 billion years ago more or less, these more complex forms of life were well installed, and the air contained more than 1 percent of oxygen. The switch from anaerobic to aerobic metabolism of oxygen-tolerant anaerobic bacteria was indeed observed to take place at this level by Pasteur. Later on, less than a billion years ago, the oxygen level reached 10 percent and an ozone layer was formed. With a shield against ultraviolet radiation in place, life could move onto dry land.

Today's air contains 21 percent oxygen. The rise of oxygen is the counterpart of the accumulation and burial of organic matter in land and sea-floor sediments. Every year, the living plants of the land extract some 100 billion tons of carbon from the atmosphere's carbon dioxide, using sunlight and water. However, the photosynthetic production of organic matter, taking place during the daylight hours of the growing season, is almost exactly balanced by respiration during nighttime and the rest of the year. The leaves fall and rot, and the process of decay, helped along by specialized aerobic bacteria, is a process of oxidation of the organic matter, returning carbon dioxide to the atmosphere. Nevertheless, some small fraction of the dead leaves and branches ends up buried by other leaves, or washed away into a river, protected one way or another from oxygen. The steady accumulation every fall and winter of this small buried fraction of the organic matter produced each spring and summer, acting over hundreds of millions of years, accounts on the one hand for fossil fuels, on the other for the oxygen accumulated in the atmosphere. In burning fossil fuels—peat, coal, oil, and natural gas—humans have vastly accelerated the rate of oxidation of the natural carbon cycle. Other rapid oxidation processes include the burning that usually goes along with deforestation, and the subsequent changes in the newly exposed soils. The different processes of oxidation are part of nature, but humans have accelerated them so much in the last two hundred years that there has been a significant and accelerating increase in the amount of carbon dioxide in the atmosphere, from 0.029 percent in 1900 to over 0.037 percent in 2000. At the same time, there has necessarily been a measured decrease in the amount of oxygen, corresponding to the oxygen used up in producing the extra CO_2. However, there is so much more O_2 than CO_2 in the atmosphere (21% compared with 0.037%) that the decrease in oxygen is nothing to worry about. The rise in CO_2 is, however, a source of worry, not as a direct threat to health, but because it provokes climate change; and as for deforestation, it certainly constitutes a problem for the diversity of life and ecological relationships even if its effect on global oxygen levels is insignificant. If all green plants were wiped off the face of the Earth, it still could take millions of years for the oxygen to disappear; but what would we (and our fellow animals) have to eat?

CHAPTER 4

CATASTROPHES

CLIMBING UP AN EVOLUTIONARY LADDER?

We humans like to think of ourselves as being on the top rung of the ladder of evolution. For those who take the Bible story literally, God created man in his image, the crown of all that He had created in His six days of labor; perhaps a bit longer, since Psalm 90:4 sings "a thousand years are but a day in Thine eyes." Even many freethinkers, neither believers in the doctrine of Creation according to Genesis, nor disciples of Pierre Teilhard de Chardin[1] or of Arthur Oncken Lovejoy and his "great chain of being," still view *homo sapiens* as sitting (or still climbing?) at the top of the evolutionary ladder, the unspecialized species best adapted to the global environment and best endowed to take advantage of any future changes. Can we regard the four billion years of evolution of life as "progress"? Prokaryotes—archeobacteria and other close relatives of the first micro-organisms—still inhabit the Earth. Forced to abandon environments exposed to oxygen, they have found ways to flourish in new environments, including some in the very innards of us "advanced" life forms. The microbes are still with us. Perhaps evolution is not a simple ladder but, rather, a web of branching paths to ever greater diversity. Indeed, since the Rio "Earth Summit" in 1992, the term *biodiversity* has become part of the language of diplomacy, and it is often argued that biodiversity is essential for the long-term survival of the biosphere, for "sustainability" (another newly fashionable word), because it ensures that there are always some species ready to adapt to any upset in the environment. Should we look on massive extinctions and reduced biodiversity due to the expansion of human activities as crimes against evolution? Or is too much emphasis being put on the notion of diversification? Even completely random evolution goes in the direction of increased diversity and complexity, just as a drunkard tends to wander further and further away from his starting point until he collapses. (I hope that women, up to now generally less well represented in this particular indignity, will not claim equality.)

If we suppose that the first living thing was the simplest possible (and not actually all that simple!), obviously all that followed had to be more complicated. However, that natural tendency in no way rules out the possibility of halts on the path of diversification, or even of backtracking to a simpler biosphere. The paleontologist Stephen Jay Gould has argued that although diversity as measured in the number of species has become enormous, *disparity*, i.e., the diversity of anatomical plans, is extremely limited. In the Canadian Rockies (Yoho National Park), the Burgess Shale layer, laid down 570 million years ago, first discovered in 1910, records the explosion of multicellular life: a plethora of extraordinary fossils, invertebrates fantastic in form, indeed one species so strange it was named *Hallucigenia*. Since the 1970s, research on these and other fossils has led to revision of the notion of ever-increasing diversity. Most families of species of the Cambrian era have disappeared; why, nobody knows. According to Gould,[2] diversification is always accompanied by decimation (in its colloquial sense of 90% destruction rather than the literal 10%). An examination of the tree of life with branches extending ever further away from the trunk (representing the assumed single common ancestor of all), reveals that many branches die out or are chopped off, never reaching the present and leaving enormous voids in the diversity of life. In the Burgess Shale, in addition to the four arthropod families that still exist today (insects, crustaceans, spiders, and scorpions), and the trilobites that disappeared much later after a long history, there are some twenty-odd families that look like nothing in the world today.

Mass Extinctions

We don't know what made the difference between those luscious branches of the tree of life that died out while others endured. It isn't only a matter of better or worse fitness to the environment, of skills for survival and for profiting from change; in the history of our planet, chance also has its role to play.[3] We humans, or rather more generally we mammals, emerged as the big winners following the catastrophe that wiped out the dinosaurs some 65 million years ago. And that was not the only catastrophe in Earth's history. Earlier, at the end of the Triassic some 208 million years ago, countless marine animals disappeared, in particular the plesiosaurs and ichthyosaurs. At that time, the extinction of many land species left free space for the dinosaurs of the Jurassic and Cretaceous. And there was an even worse *hecatomb* (from the Greek, literally a sacrifice of a hundred oxen, but used to mean much worse) earlier still, some 250 million years ago, at the conclusion of the Permian in the Paleozoic. Many plant and animal species had established themselves on the land long before those days. During the Carboniferous era, for tens of millions of years, dense swamp forests (in particular trees reproducing by spores, of the genera *Lepidodendron* and *Sigillaria*) stored up colossal amounts of carbon destined to be the coal reserves being burnt up since the start of the Industrial Revolution. Insects swarmed over the land, with some exceptional species such as the giant dragonflies (wingspread over 18 inches!) of which fossils have been found in France. Many families of amphibians and reptiles also made their appearance. And then, in only a few million years, between two-thirds and 80 percent of these families became extinct. Even the insects, usually the great survivors of evolution's disasters, did not escape the holocaust: 30 percent of the insect orders disappeared, perhaps because of the loss of plant species. What caused such a catastrophe? One line of investigation homes in on a relentless drying out of the continents, a calamity for plant life with a long chain of disastrous repercussions for insects, amphibians, and reptiles. But what caused this desiccation? When the changes began, there seems to have been a significant drop in sea level, reducing the area supplying water by evaporation to the atmosphere and increasing the land area over which reduced rainfall had to be shared, leaving high and dry the wetlands along the coast and in estuaries, usually abodes abundant in life. But such an explanation only raises another question: what caused the sea level to fall?

The mass extinctions at the end of the Permian also decimated the marine domain. Douglas Erwin[4] of the Smithsonian Institution recounts that before that time, an "exquisite" marine fauna existed, mostly composed of immobile forms—corals, echinoderms, and other animals attached to the sea bottom—but with very few mobile species such as fish, cephalopods (squid, octopus), snails, and other gastropods. Whatever the catastrophe was, it hit most strongly those communities living in shallow water or around coral reefs; and indeed a rapid fall in sea level leaving them high and dry could account for that. Only one genus of sea urchins survived, but it has done very well since then and can be found everywhere.

In southern China, rocks dating from the end of the Permian contain layers of ash indicating strong volcanic activity during a period of less than a million years—very short, geologically speaking. According to Yin Hong-fu and his colleagues of the University of Geosciences of China, more than 90 percent of invertebrate species disappeared during this Changxi stage. Overall, between 80 and 95 percent of marine species became extinct. It took marine life several million years to reestablish itself in the seas after this disaster, and the new seascape was radically different: many more mobile fauna, predators, fish, snails, and crabs. Since then, the survival of potential prey depends on special conditions or the existence of possible hiding places in rocks, reefs, or in the silt at the bottom of the sea.

Complete understanding of the processes involved in the Permian extinction is still lacking. What was the impact of the frenzy of volcanic activity and of the climate changes that it could have caused? Did the drop in sea level intensify erosion at the edges of the continents, oxidizing large masses of organic matter with strong release of carbon dioxide, there again perturbing the climate and maybe even reducing atmospheric oxygen? Such changes would certainly have modified ocean chemistry, and there are indications that in some places there were large pockets of ocean water completely devoid of dissolved oxygen. It all comes back to the puzzle of the sudden drop in sea level and, to complicate things still further, its equally puzzling rise, nearly as sudden, only a few million years later. Leaving aside science fiction fantasies of the ocean suddenly disappearing into chasms leading to the Earth's mantle, can one nonetheless imagine some sort of spasm of plate tectonics?

THE DEMISE OF THE DINOSAURS

More is known about the disaster that did in the dinosaurs, clearing the terrestrial stage for the multiplication of mammals. Fish continued to swim the oceans, and insects, as always, survived. Maybe it should be said that rather than being blown away, the dinosaurs flew away; according to some paleontologists at least, birds descend (or rather ascend!) from dinosaurs. Geologists and paleontologists define the boundary between layers of the Mesozoic and the Cenozoic eras, of the Cretaceous and the Paleocene periods, according to the changes in the types of rocks and fossils they contain. Before (below) this limit 65 million years ago, the strata are rich in calcareous sediments, accumulations of fossils of nannoplankton (microscopic algae) deposited on the sea floor. If you take the ferry rather than the tunnel to get to England from France, you can see them along both chalky coasts of the English Channel (or La Manche, in French), and on a clear day you can see the white cliffs of Dover from the upper deck even before leaving Calais. Above the 65-million-year limit, sea-floor sediments contain much less calcium carbonate, and fossils of several families of mollusks are no longer found. And in continental sediments, frequently containing dinosaur fossils below (before) 65 million years ago, none can be found above. By contrast, new families of mammals appear, including large mammals for the first time.

The event that put an end to the reign of the dinosaurs and the Mesozoic era, giving mammals their chance, was a worldwide ecological catastrophe by all counts: nearly half of all *genera*[5] disappeared. Although the mass extinction of the Permian period about 250 million years ago was worse, the disappearance of the dinosaurs strikes the imagination. Many of us have viewed their impressive skeletons in museums, most of us have seen drawings in books, not only schoolbooks, and of course Stephen Spielberg had great success with his extraordinary dinosaur animations in *Jurassic Park* (1993). It is hard to imagine, except in nightmares, that such enormously powerful creatures could have been wiped out. Scientists wondered for many years about what could have caused their rapid disappearance, coming up with a great variety of theories and scenarios. For some, it could have been due to unfavorable mutations triggered by a dramatic increase—by a factor 10, 100, 1,000?—in cosmic-ray particles reaching the Earth after a supernova explosion somewhere in the neighborhood of the solar system. For these high-energy particles to affect life, they would have to get through the protective barrier of the Earth's magnetosphere. That could have happened if the cloud of particles from the supernova explosion reached the Earth during a period when the magnetosphere was weakened, something that may happen when the Earth's magnetic field flips. And we know that the magnetic north and south poles of the Earth switch on the average twice every million years. However, this is not the only possible explanation for dinosaur destruction. Other theories have raised the possibility of strong climate changes in the Tropics (but they then must be explained). Certainly, if climate changes, the changed distributions of temperature and rainfall modify the conditions that favor one ecosystem over another. The extinction of a particular family, genus, or species may result from a complicated chain of indirect causes and effects. Over thirty years ago, Carl Sagan quoted one suggestion that the demise of the dinosaurs resulted from constipation after the disappearance a particular laxative species of fern.[6] Other theories involved a worldwide cold wave following the spread of a layer of cold but not very salty water over the world's oceans, floating on the surface because, with its low salinity, the water was less dense.

Two theories remain in competition today. Some specialists argue that the extinction of the dinosaurs corresponds to a period of intense volcanic activity. It's not a question of just one or even of a thousand eruptions comparable to the explosion of Krakatoa (Indonesia) in 1883, but rather of a prolonged period of activity, long compared with the thousands of years it takes for a complete turnover of the oceans, but short from the geological point of view. On the Deccan plateau in India, basalt rocks cover more than 500,000 square kilometers (nearly 200,000 sq. mi.), and correspond to massive lava outflows occurring precisely at the end of the Cretaceous, 68 million years ago.[7] This sort of outflow could correspond to volcanic activity similar to the activity that drives sea-floor spreading, with lava emerging from elongated fractures in the crust rather than from craters. The volcanic convulsion that buried the Deccan plateau in lava must also have changed the composition of the atmosphere and severely affected climate. Initially, there must have been strong sudden cooling resulting from the blocking of sunlight by the sulfate aerosol veils in the stratosphere.[8] If strong cooling lasted a year after the formation of the aerosols, it would have been the death of tropical species unable to adapt to such a "volcanic winter." However, a long period of strong volcanic activity (again, remember, thousands of Krakatoas) would at the same time have added a substantial amount of carbon dioxide to the atmosphere, reinforcing the greenhouse effect. This would gradually warm things up, ending the extended cold snap and producing global warming together with geographic shifts of humid and arid zones.

Certainly enough things would change to upset living conditions, leading to the extinction of some species while others profited, if only from the disappearance of predators.

DEADLY IMPACT

One of the peculiarities of the Cretaceous-Paleocene transition, marked in 65-million-year-old rock, is the existence of a thin layer enriched in the rare metal iridium. This anomaly, first noted in 1980 as an enrichment by a factor 30 in clay near Gubbio in Italy, occurs in a layer less than 2 cm thick (about ¾ in.), corresponding to an extremely short time interval. Below the layer, calcareous nannoplankton fossils are abundant; above it, hardly any are found. Since 1980, the iridium-enriched layer has been identified in nearly all strata containing the Cretaceous-Paleocene transition, with the enrichment factor ranging from 20 to 200, and always with the same age of 65 million years, estimated using radioactive isotope dating. Where does the iridium come from? Today, most specialists have accepted as correct the hypothesis proposed by the California team led by physicists Luis and Walter Alvarez in 1980: the iridium comes from the impact of a meteorite or comet. Certain classes of meteorites are known to be relatively rich in iridium, and the quantity of iridium in a body 10 to 20 km (6–12 mi.) in diameter would be enough to account for the extra iridium in the Cretaceous-Paleocene boundary layer over the entire globe. Does the discovery of the iridium anomaly shoot down all the other theories of dinosaur extinction? Some geophysicists argue that strong prolonged volcanic activity could also eject extra iridium into the atmosphere and ultimately spread it over the Earth's surface. But they are the minority: nearly all geoscientists accept that the extinction of the dinosaurs, and with them of about half of the biosphere, resulted from the impact of a large meteorite or small comet. Furthermore, thanks to satellite imagery and measurements made by offshore-oil prospectors, the impact crater has finally been found, near the town of Chicxulub on the Yucatan Peninsula of Mexico. The crater, more than 150 km (about 100 mi.) across, extends partly over the sea floor off the present-day coastline, and this made it hard to recognize.

The impact and explosion must have had truly apocalyptic results. Within a few hours, earthquakes must have rocked many regions of the globe, with giant tsunamis sweeping over coastal plains. Some of the impact and explosion debris fell far and wide within days of the impact, burying some areas in thick layers of dust enriched in iridium, the dust later being packed down into the thin layer that still marks the event today. Part of the debris must have reached the stratosphere, spreading out over weeks and months into a dust and aerosol layer blocking the Sun. A long night fell on the Earth and, just as for a volcanic spasm, the long and out-of-rhythm winter put an end to photosynthesis and plant growth for months and years, freezing the land. In the Tropics, very little survived, and even in the temperate and polar zones where plant and animal species are well adapted to the annual return of winter, the prolonged lack of sunlight and warmth put the biosphere to a severe test. There is no way to know during what season the event took place, but over half of the Earth it must have been spring or summer, the worst time for the Sun to disappear. In the oceans, except for the very special biological communities living off the hydrothermal activity along the mid-ocean ridges, the blocking of sunlight caused a catastrophe starting with phytoplankton and propagating along the food chain, even though the sea did not cool off as quickly as the land. The rest of the story must have been more complicated. The area where the impact occurred is and was very rich in calcareous sediments, so that the impact must have released enormous quantities of carbon dioxide, adding to the greenhouse effect, producing centuries or more of global warming after the deep freeze ended. Some species may have managed to survive the first climate shock only to fall victims of the second phase.

Would we humans, or mammals in general, have had the success we have had without the destruction of the dinosaurs? This was neither the first nor the last major impact on our planet, but it was certainly the most recent one with worldwide consequences. However, marks of more than a hundred major impacts have been identified, some of them quite recent. In Arizona, tourists can visit the Barringer Meteorite Crater, nearly a mile wide, 500 feet deep, formed some 50,000 years ago when a metallic asteroid less than 90 feet in diameter plunged into the desert. In the twentieth century, sizable meteorites (or small comets) twice hit Siberia, in 1908 and in 1947, flattening large areas of forest and frightening observers in the taiga, probably perturbing the upper atmosphere as well but not having any detectable influence on climate. For more ancient impacts, recognition of the iridium anomaly as an indicator has provided a new research tool.

The identification of an iridium anomaly in a core of sea-bottom sediments, obtained by the U.S. research vessel *Eltanin* from the floor of the Antarctic Ocean west of Cape Horn, has revealed a mid-ocean asteroid impact, not 65 but only 2.15 million years ago, at the end of the Pliocene. Seafloor spreading has had little time to move the material, so that interpretation of the acoustic soundings and of the cores obtained by deep-sea drilling is fairly straightforward. Exploration of the site by the German research vessel *Polarstern* (*Pole Star*) has shown that the debris are spread over a few hundred miles at a depth of 4,700 meters (2,570 fm). A fragment of the meteorite was recovered, and it turns out to be a relatively rare type, a *mesosiderite*. The debris include nickel-rich spherules and other minerals also found in the layer resulting from the Chicxulub impact. However, the *Eltanin* meteorite does not seem to have penetrated to the deep ocean bottom. From the nature and distribution of the debris, it must have been between 1 and 4 kilometers (0.6 and 2.4 mi.) in diameter; energy released by the impact and explosion must have been equivalent to a million tons of TNT. Some explosion fragments could have reached anywhere on the Earth, and others could well have formed a stratospheric dust and aerosol veil causing sudden cooling of which some trace exists. Nevertheless, there does not seem to have been any major impact on the biosphere. A narrow escape for our ancestors! However, such an impact in the ocean made a big splash, to say the least. Calculations show that it must have produced a "mega-tsunami" with waves over a hundred feet high in mid-ocean, possibly becoming monster tidal waves on reaching the continental coast, washing up to altitudes as high as hundreds of meters (over a thousand feet). In Pisco (Peru), coastal sediments dating from this period contain a suggestive mix of marine and terrestrial animal fossils.

There must have been many such impacts; a meteorite a kilometer (over 3,000 ft.) in diameter probably hits the Earth once every million years on average. Although that is not a monster comparable with the Chicxulub impact that wiped out the dinosaurs, it can still wreak havoc along all coasts if it falls into the ocean, or demolish a medium-sized country if it falls on land. Some astronomers believe that such impacts occur mainly when the solar system passes through clouds of cold dark objects as it moves around the center of the Galaxy. This would define a cycle with periods of great danger separated by longer and more peaceful intervals lasting 20 to 30 million years. It has even been suggested that *all* of the mass extinction events of the last 250 million years were caused by meteorite or comet impacts, but this hypothesis is not generally accepted. There are many other possible sources of disturbance of the environment, and complex interactions between different families of species can sometimes amplify even apparently minor perturbations. Nevertheless, it must be recognized that potentially dangerous objects lurk in the neighborhood of the solar system. Should we worry about them, or would that just be another way of fearing that the sky is falling? A number of astronomers argue that we should watch the skies more carefully, and make a thorough inventory of *all* dangerous objects, such as asteroids larger than, say, 100 meters or yards across, moving in orbits that could one day intersect the orbit of the Earth. And suppose we discover some such threat, what then? According to Tom Gehrels, in charge of the Spacewatch program at the Kitt Peak National Observatory in Arizona, with less than fifty years' warning, nothing much could be done to avoid disaster: all that would be left to us would be to visit the places we love most and say goodbye to our friends. Others are less pessimistic regarding our ability to react rapidly and effectively. With enough time (but how much would be enough?), we could protect Earth by nudging the deadly object into a different trajectory, if necessary using thermonuclear-armed missiles. With more than a century's forewarning, missiles carrying conventional explosives would suffice. You may think that's all science fiction, but the very serious British weekly magazine *The Economist* featured an article on this in its centennial issue (not April 1st!) several years ago. Of course, such a possible protective application could be used as an argument to justify maintenance of part of the nuclear arsenal accumulated during the Cold War. And some cynics or alarmists have pointed out that if the orbit of an asteroid can be perturbed so that it misses the Earth, terrorists (with rockets and bombs) could threaten to push an innocent object *into* a doomsday orbit. Nothing's simple.

MAN THE EXTERMINATOR

Sixty-five million years after the catastrophic termination of the Cretaceous, are we humans the new exterminators? Although the Chicxulub impact was in itself a spectacular instantaneous cataclysm, the disappearance of the dinosaurs and of many other life forms dragged out over at least a million years. By comparison, during the four centuries that have gone by since 1600, over 1 percent of bird and mammalian species have disappeared. Today, the rate of extinction is dozens or hundreds of times its average in the past, and the extinction rate is now probably as high as it was during the catastrophic Cretaceous-Paleocene transition. No need to look for cosmic invaders to explain this dramatic increase of the rate of extinctions: we humans are responsible. Nor is this something new in terms of human history. Many species of large mammals disappeared only 10,000 years ago, at least in part as a result of overhunting by humans. There seems to have been a wave of extinction from the Bering Strait down to the tip of South America, following the advance of the peopling of the Americas. No doubt climate change and the retreat of the ice caps played a major role, but hunting by humans must also have hastened the disappearance of the mammoths in Eurasia. With more and more grazing domesticated animals and with spreading agriculture, less and less land is left for the natural biosphere, and overall there are fewer living species on our planet. Humans have long sought to control water resources on land, and agriculture and industry are changing the global atmosphere. Some see all this as a catastrophe in midcourse.

CHAPTER 5

ICE, MOON, AND PLANETS

THE MOON, TIME, AND TIDE

Cosmic catastrophes, whatever they may be—nearby supernova explosions, solar "hurricanes," or comet impacts—won't stop the world going round. The Earth spins at over 1,000 miles per hour at the equator, 700 mph at latitude 45°N of Minneapolis. Should the Earth's rotation suddenly come to halt, everything not tied down to solid ground—atmosphere, oceans, cows, cats and dogs—would be whipped away, as described vividly by H. G. Wells.[1] In 1950, self-taught Egyptologist Immanuel Velikovsky told a lively tale about how biblical miracles could be explained by the birth and expulsion of a massive body from the planet Jupiter, passing next to the Earth several times to help part the Red (or Reed) Sea and carry out other sundry good works, and stopping the Earth's rotation just in time to make the walls of Jericho come tumbling down and give victory to Joshua,[2] before finally settling down as the planet Venus, the evening (or morning) star. Velikovsky acquired a faithful following and struck it rich with his best-selling *Worlds in Collision*,[3] but he never convinced any serious scientists, even though he did write—correctly as it turned out, but for the wrong reasons—that the surface of Venus had to be infernally hot. Certainly there is no way to bring the Earth's rotation to a sudden stop without at the same time causing absolutely cataclysmic damage to the planet as a whole. But that does not mean that the Earth has always turned at the same rate; in fact, it *has* slowed down over the ages. The daily and annual growth rhythms of certain types of corals are recorded in calcareous layers. In contrast to today's 365 days and a bit more per year, coral fossils of the Devonian period, 370 million years ago, reveal about 400 days per annual cycle. The Earth then turned in only 22 hours.

This braking of the Earth's rotation continues, increasing the day's length by about 1.6 seconds every 100,000 years. Thanks to their extremely precise measurements, astronomers detected this already in the nineteenth century. This was part of their responsibility for keeping track of time, publishing the nautical almanacs and other tables needed for celestial navigation.[4] The rotation of the Earth with respect to the stars takes 24 hours of sidereal time, i.e., 23 hours, 56 minutes, 4-*plus* seconds of mean solar time; the difference corresponds to the motion of the Earth around the Sun. But the planet's rotation is slightly irregular, and with better and better measurements (but not predictions) of these irregularities, the rotation of the Earth no longer provides the best standard of time, either for the equations of celestial mechanics (*ephemeris* time) governing the motions of the planets, or more generally for physics (*atomic* time, as in "atomic clocks"). A further complication arises from the need to have a whole number of days—365, with 366 in leap years—in the year of the seasons corresponding to the motion of the Earth around the Sun, so that every so often an additional second is inserted, a 61st second in a particular minute. That needs to be taken into account when programming the operation of an Earth observation satellite traveling at 8 kilometers (about 5 mi.) per second.

It is the tides that slow down the Earth's rotation. Tides have been known to humans for as long as they have inhabited ocean shores, but the understanding of their cause came with Newton's law of universal gravitation. The mass of the Moon attracts masses on the Earth, just as the Earth attracts the Moon and keeps it in its orbit. But the attraction is stronger when the distance is less. Masses on the side of the Earth facing the Moon are more strongly attracted than the average for the Earth; masses on the opposite side less strongly. This applies in particular to the easily distorted fluid envelopes of our planet: the oceans (liquid) and the atmosphere (gaseous). As a result, a planetary-scale double wave is formed, with one crest on the side of the Earth facing the Moon, the other crest on the opposite side. The wave follows the Moon as it moves around the Earth, but at the same time the solid Earth is spinning about thirty times faster. For an observer standing at the seashore, on the spinning solid Earth, two wave crests are observed in a period of 24 hours 50 minutes, the interval from one high tide to the next being 12 hours 25 minutes on average. The fact that the Sun's gravitational attraction also raises tides complicates the picture somewhat. However, although the Sun is much more massive than the Moon (by a factor 2,676,000), lunar tides are stronger because the Moon is so much closer (389 times closer) to the Earth; the tide-raising forces depend in fact on the *differences* in the gravitational forces acting on the different parts of our planet.[5] The strongest tides ("spring" tides) occur (not only in spring) when the Moon is either new or full, Sun and Moon then lying on a straight line going through the Earth's center. This configuration is known as *syzygy*, to use a technical word otherwise encountered in spelling bees and crossword puzzles. If this happens at the vernal (spring) or autumnal equinox, the alignment is perfect and the tides are extremely strong, with very high high tides causing floods and very low low tides making navigation difficult. By contrast, when Sun and Moon are in quadrature (first and third quarter Moon), their tide-raising forces partly cancel, giving the weak neap tides, with high tide not very high, low tide not very low.

Sailors of ancient times, or at least those who sailed along coasts where tides are fairly strong (along the Atlantic, for example, rather than the Mediterranean), certainly must have realized that tides follow the Moon and its phases, but the physical explanation had to wait for Newton and his laws.[6] In addition, the details of the tides at a particular place depend not only on the positions of the Moon and Sun in the sky but also on the shape of the ocean basin, the contour of the seashore, and the slope of the seabed. With careful analysis of the observations, reliable tide tables can be constructed, giving for example the times of high and low tide for each day at Boston harbor, together with corrections to be applied to get the times up and down the Massachusetts coast at Plymouth, Marblehead, and Gloucester. It still is very hard to calculate the detailed timing and strength of tides on a purely theoretical basis, and to understand completely the details of what is observed. In fact, only since 1992 have observations of tides in mid-ocean been possible, using radar altimeters installed onboard artificial satellites (such as the French-American Topex-Poseidon mission) whose orbits are also very precisely determined. Along continental and island coasts, tide gauges provide direct measurements, and some of them have been in operation for many years.[7] Although, as a rule, two high tides and two low tides come on average every 24 hours and 50 minutes, around some islands in the Pacific there is only one tidal oscillation in this period. Ebb and flood are barely discernible in the Mediterranean, but in other seas the amplitude is enormous: almost 50 feet between low and high tides in the Bay of Mont-Saint-Michel in France, even more in the Bay of Fundy between Nova Scotia and New Brunswick, or in nearby Passamaquoddy Bay between Maine and New Brunswick. When high tide comes in at Mont-Saint-Michel, it can outrun a galloping horse. And when the flood tide enters an estuary, it can in some places and at some times raise an abrupt jump in the water level (called a "mascaret" in the Seine River estuary above Le Havre), a swiftly advancing dangerous wall of water. In the Quangtang Jiang estuary in southern China, it can be nearly thirty feet high. It's dangerous to be on the water when it comes.

The global tidal bulge tends to follow the Moon and the Sun; where it reaches the coast or the edge of the continental shelf, its "push" in effect resists the Earth's more rapid rotation, and this tends to slow down the Earth's rotation. This friction between solid Earth and liquid water is increased when a power plant is installed to harness tidal energy, as in particular in the Rance estuary in Brittany (western France);[8] the increase is slight on the planetary scale, and would be so even for a power plant in the Bay of Fundy. Tidal power plants can create ecological problems locally without helping significantly to satisfy the need or greed for energy worldwide. At any rate, putting braking action on the rotation of the solid Earth, the tidal bulge in the oceans is subjected to an equal and opposite reaction, giving it a slight advance relative to the positions of the Moon and Sun. This advanced bulge exerts a small but significant gravitational attraction on the Moon, pulling it ahead in its orbit. Over very long periods of time, the Earth's rotation has slowed, and the transfer of angular momentum to the Moon has gradually pushed its orbit further outward. Long ago, Earth and Moon were much closer together than today. And in a distant future, in billions of years, with the Earth's rotation slowing and the Moon drifting further and further away, the month and the day will finally have the same length, 47 of our present days.

THE PRECESSION OF THE EQUINOXES

In the year 129 B.C., long before anyone had discerned the slowing of the rotation of the Earth (or as most saw it then, of the sky), the great astronomer Hipparchus of Nicaea, of the school of Alexandria (in Egypt, but under strong Greek influence then), discovered or at least described the precession of the equinoxes. It's thought that astronomers, in China, India, perhaps also in Babylon, discovered the phenomenon independently. The lunar-solar precession, like the tides, is an effect of the gravitational attractions exerted on our planet by the Moon and the Sun, and it could not be understood before Newton worked out the laws of motion and of gravity. Let's look into this. As we all learned in school, the seasons result from the fact that the Earth's axis of rotation is tilted relative to the axis of the Earth's motion around the Sun, so that the plane of the Earth's equator does not coincide with the *ecliptic*, a term designating (in modern terms) the plane of the Earth's orbit around the Sun, corresponding to the apparent motion of the Sun in the course of the year. The angle of 23°27' between the two planes is called the *obliquity of the ecliptic*. Starting with the spring equinox (March 21) and through northern summer, the Sun's path lies north of the celestial equator (the projection in the sky of the Earth's equator), and it rises higher in the sky, stays longer above the horizon, and generally illuminates and heats the Northern Hemisphere more than the Southern. The rest of the year, the Sun's apparent path takes it south of the celestial equator, and the situation is reversed: it's autumn and then winter in the Northern Hemisphere, and it's the turn of those who live south of the equator to enjoy spring and summer. The apparent path of the Sun along the ecliptic crosses the celestial equator in two points corresponding to the equinoxes (the days when day and night are of the same length— twelve hours— all over the Earth), from south to north at northern spring equinox in March, at the "First Point of Aries," a point given the symbol ♈ representing the horns of Aries the Ram. Although it's only during a total eclipse of the Sun that the stellar background of the Sun's position can be physically seen, ancient astronomers could easily imagine the path of the Sun along the ecliptic, against the background of the constellations of the zodiac which mark that path (and not only ancient astronomers, but you too).

Comparing his own observations to those made by other astronomers 144 years before, Hipparchus found that the angular separation between the point ♈ and the star Spica (α Virgo) had increased by two degrees. In fact, the point ♈ shifts relative to the background of stars by 50 arc seconds per year, so that it makes a complete tour of the sky in about 26,000 years. That should raise a few questions for those who continue to consult their horoscopes and attach credence to what Saint Augustine called the "ridiculous prophecyings and blasphemous follies of the astrologers."[9] The signs of the zodiac did indeed still coincide with the constellations of the zodiac when Hipparchus lived, or in Augustine's time 500 years later; but today, ♈, the First Point of Aries, defining the zodiac *sign* Aries, no longer lies in the *constellation* Aries; it has shifted into the constellation Pisces. And "soon" it will enter the constellation Aquarius, inaugurating "The Age of Aquarius" (a great song of the 1960s). But therein lies a dilemma: should astrology be based on the *signs* of the zodiac, which at least correspond to the seasons and so are related to changes in our terrestrial environment? Or should it be based on the constellations? And why should the properties of Aries the Ram, which some observers in some civilizations more or less imagined seeing in the arrangement of stars which formed the background of the Sun at the beginning of northern spring twenty-two centuries ago, have anything to do with the prospects of a baby born in March 2001, when those same stars are no longer even behind the Sun?

To visualize the precession of the equinoxes, consider the orientation of the Earth's axis of rotation, passing through the North and South poles. Today, inhabitants of the Northern Hemisphere are lucky to have a fairly bright star quite close to the celestial north pole, marking the direction of north on nights when the sky is clear: the Pole Star or *Polaris* (α *Ursae Minoris*), brightest star in the constellation of the "little bear" (or the Little Dipper). But Polaris was no pole star for the Babylonians 5,000 years ago; the stars in the sky then appeared to move around the star α *Draconis*. There will be no bright star to mark the North Celestial Pole 2,000 years from now; but in 11,800 years, our descendants, if any are around, will have Vega (α *Lyrae*), a beautiful star of our present summer nights, next to the pole. What's happening is that the Earth's axis of rotation moves around a cone in about 26,000 years, just as the axis of a rapidly spinning top wobbles in a couple of seconds. The spinning motion of the top tends to keep it erect while gravity is "trying" to pull it down, and the conical wobble results. The

Earth is not a perfect sphere, having an equatorial bulge precisely because it is spinning. The gravitational attraction exerted by Moon and Sun on the Earth, acting on average in the ecliptic plane, "tries" to pull the equatorial bulge "down" to the ecliptic, "tries" to make the Earth's axis of rotation perpendicular to the ecliptic. But just as the spinning top does not fall down immediately (until friction slows its spin), the gyroscopic effect of the Earth's spin transforms the gravitational pull toward the ecliptic into the slow conical motion of the Earth's axis. The Earth's shape, its spin, and the precession of the equinoxes depend on one another. Isaac Newton worked out the equation expressing these relationships, and he used it to calculate the Earth's equatorial bulge. Newton's estimate was confirmed by measurements of the shape of the Earth near the Arctic Circle, made by a French expedition to Lapland in 1736–37.[10] After a quarrel, French philosopher Voltaire could not resist the temptation to make fun of expedition leader Maupertuis and his achievement:

> O Heroes of Physics, like Argonauts of old,
> Scaling high peaks, on seas a-sailing bold,
> Returning from those climes ruled by the Swedish crown,
> With kit, compass, and your lives, and also two Lapp wives.
> From that awful desolation, you've brought back confirmation,
> Of what great Newton knew without leaving his home station.
> You've succeeded in surveying a part of little worth
> Of the ever-frozen flanks of the slightly flattened Earth.[11]

THE MOON AND MILANKOVITCH

The science of celestial mechanics, based on Newton's universal laws and developed by Pierre Simon Laplace (1749–1827), provides the framework for understanding and predicting the irregularities of the Earth's rotation and the motions of the Moon and planets. Now those may appear to be details, but in fact, the exact values of these parameters govern the existence of life on our planet.[12] And these parameters are not fixed; they vary as a result of the complex interplay of the gravitational pull of the Moon, the Sun, and also other planets, on the Earth. In the nineteenth century, as geologists began to realize that climate had gone through enormous changes in the distant past (in fact, not so distant), some scientists—Joseph Adhémar in France, James Croll in Great Britain—began to develop the idea that these climate changes must have resulted from the astronomical parameter variations.

In 1911, Milutin Milankovitch, a young Serbian mathematician looking for examples to include in his course on mathematical physics, came across the following problem stated by the French Academy of Sciences in 1885: "Distribution of heat on the Earth's surface. Determine theoretically the law governing the distribution of solar heat input at different latitudes in the course of the year, taking into account atmospheric absorption." Inspired by this problem, Milankovitch set himself a more ambitious target: "The equations for the solar energy input to the Earth should be written in a form making it possible to calculate long-term variations, so that the paleoclimate problem and in particular the problem of the ice ages could be attacked."[13] World War I interrupted Milankovitch's work in 1914. As he told it: "Although a prisoner of the Austro-Hungarians, I was allowed to leave the prisoner-of-war camp at the end of 1914 to go to Budapest, where thanks to the hospitality of the Hungarian Academy of Sciences, I was able to complete the work under way." Thanks to this survival of civilization in the midst of the horrors of war, Milankovitch was able to elaborate his theory, which still today guides thinking on variations of climate of the past few million years.

At present, the Earth goes through perihelion, the point of its orbit closest to the Sun, early in January, when it's winter in the Northern Hemisphere; in July the Earth is at its greatest distance from the Sun, at aphelion. The orbit is not a perfect circle; its fairly weak eccentricity of 0.017 produces a modulation of ±3.4 percent in the flux of solar radiation reaching the Earth (above the atmosphere). From a maximum of 1,415 watts per square meter (131.5 W or watts per square foot) in January, solar radiation flux falls to 1,320 W/m^2 in July, 7 percent less. This tends to reduce seasonal contrasts in the Northern Hemisphere, to increase them in the Southern. In fact, however, seasonal contrasts are generally stronger in the north than they are in the south. This contradiction shows that astronomy does not determine everything on Earth. The two hemispheres of our planet differ principally in the proportion of land and sea. As solar radiation input increases, land surfaces warm up more quickly than do the oceans. However, if (when) Northern Hemisphere winter coincides with aphelion, when solar flux is at its lowest, winter temperatures will be much lower than now. That situation recurs regularly; the last time was about 11,000 years ago.

The precession of the equinoxes corresponds to the gradual drift of the directions of the equinox and solstice points relative to the directions of the stars, with a complete tour in 26,000 years. However, it is the Sun, and no other star, that determines terrestrial climates. In fact, the orientation of the slightly elliptical orbit of the Earth and the directions of perihelion and aphelion are also changing relative to the stars. This precession takes 70,000 years for the grand tour, and it results from the perturbations of the Earth's orbit by the gravitational attraction of the other planets, mostly of Jupiter, the biggest. The upshot of all this is that the equinox and solstice positions shift relative to the positions of perihelion and aphelion with a complete tour in about 21,000 years. This cycle, and not the 26,000-year cycle, is the one that counts for climate.

But that is not the end of it. Seasonal contrasts depend in the first instance on the obliquity of the ecliptic. For zero obliquity, there would be no seasons apart from a weak global variation, the same both north and south, resulting from the variation in the Earth-Sun distance. For obliquity close to 90°, as is the case for the planet Uranus, the contrasts would be spectacular. In Paris or in Seattle, winter would involve ten weeks of night as near the poles today, while a midnight Sun would reign over summer. There would be "normal" rising and setting of the Sun only in spring and fall. Is there any risk of this happening? The gravitational pulls of the other planets do slightly perturb the obliquity of the ecliptic; it varies between 21° and 25° with a period of about 40,000 years. By contrast, although at present the situation of Mars, with an obliquity of 24°, resembles the Earth's, its obliquity may have varied enormously in the past, and this would explain the strong climate changes implied by the traces of past existence of liquid water, and even of an ocean, on the presently frozen and dry red planet.[14] Why such big variations on Mars, and not on the Earth? The Earth has a massive Moon, but Mars has only two minuscule natural satellites, Phobos and Deimos, their masses much too small to have any effect on the planet's axis of rotation. Of course, Mars *is* subject to the gravitational attractions of the other planets, in particular Jupiter and the Earth, and over a few million years the cumulative effect of these perturbations can provoke huge chaotic variations in the Mars obliquity. Such variations have not occurred on Earth because of the relatively strong gravitational pull of our Moon on the Earth's equatorial bulge, acting to stabilize the orientation of the Earth's axis, thus limiting the fluctuations of the obliquity of the ecliptic to only a few degrees.[15] If this is correct, we need not fear a ten-week-long winter night in Seattle or Paris; our faithful companion the Moon keeps obliquity chaos from our skies.

Weak as they may be, the obliquity fluctuations significantly influence the seasonal rhythms of sunshine at high latitudes. In addition, the gravitational perturbations by the other planets induce fluctuations in the shape of the Earth's orbit around the Sun: its eccentricity varies between practically zero (the value for a perfect circle, for which case the flux of solar radiation reaching the Earth would remain perfectly constant throughout the year), and a maximum value of 0.045 for which incident solar radiation flux oscillates by plus and minus 9 percent. Calculations show that eccentricity must vary with periods of 100,000 and 400,000 years. Combining the quasi-periodical variations (21,000, 40,000, 100,000, and 400,000 years) of the astronomical parameters (precession, obliquity, eccentricity) gives a complicated pattern of modulation of incoming solar radiation. Comparing different epochs and assuming that the Sun itself remains constant, *changes* in summer insolation at latitude 65° North can reach 60 watts per square meter, enough to decide whether winter snows completely melt over the summer, or whether some remain.

ICE ON THE MOVE

How exactly do the astronomical variations affect climate? Do ice sheets tend to spread during epochs of long, cold winters? Or, on the contrary, when winters are relatively mild but with heavy snowfall?—especially when accompanied by relatively cool summers during which the accumulated snow and ice does not completely disappear. Milankovitch continued working out details of his theory until 1940, but it was not accepted by most climatologists for many years. However, with improved calculations of planetary mechanics, and even more important, with the birth of modern paleoclimatology—the science of ancient climates—his ideas were finally taken seriously.[16] Thanks to the development of isotopic analyses, precise dating of geological layers became possible, and a way was found to take ancient temperatures. Annual layers can be identified and counted in the layers (called *varves*) of ancient lake beds, resulting from the seasonal cycle of sediment deposition. Annual snowfall layers can also be counted in the ice cores extracted from the Greenland and Antarctic ice caps, and from mountain glaciers. The development of radioactive isotope analysis revolutionized dating, making it possible to check that nothing was missed in a count, and to find the age in years of seabed sediments, and not just the order in which they were deposited. Such dating uses the "clock" of the naturally radioactive isotopes carbon-14, thorium-230, protactinium-231, uranium-234 and -235, potassium-40, and others. Each of these has its characteristic "half-life," a period in which half of the isotopes in a particular sample will have decayed into something different and distinguishable. Other isotopes serve to estimate past temperatures, using the fact that changes of water between the solid, liquid, and gaseous states depend on the temperature and on the exact mass of the H_2O molecule, and that there exist different isotopes of hydrogen and oxygen in nature: mainly ordinary hydrogen-1 and deuterium (hydrogen-2) as well as radioactive tritium (H-3), and ordinary oxygen-16 and O-18. Depending on which of these isotopes go into the H_2O molecule, it is more or less massive, and this affects the changes of state. With a precise determination of the isotope ratios in a sample of ice from the past, one can estimate the temperature of formation of the snowflake that later was transformed into ice.

One can also estimate how much of Earth's water was frozen at different moments in the past, a fraction that has often varied. The last ice age or Last Glacial Maximum (LGM) peaked only 18,000 years ago (18,000 B.P. or "Before [the] Present" as the specialists put it, with the present taken by definition to be 1950). At LGM, sea level was some 120 meters (about 400 ft.) lower than today, because about 50 million more cubic kilometers of water (originally from the sea) were sitting in solid form (ice) on the land, in northern Europe and the northeast of North America as well as in thicker Greenland and Antarctic ice caps. The sea-level *regression* exposed presently submerged continental shelves to the air, and animals including humans could cross on freezing feet between England and France, between eastern Siberia and Alaska. Many rivers could reach the lower sea only by carving valleys or canyons in the continental shelf, now flooded, as for example the Hudson River Canyon extending a few hundred miles beyond present-day New York Bay. Stories of a flood are found not only in the Bible but also in the Gilgamesh epic and other legends of the Near East, of India, and still other places. Some scholars believe that these refer to a sort of folk memory of a relatively rapid sea-level rise that must have taken place several thousand years ago. In the Nile valley, on the shores of the Persian Gulf, and along many other coasts and estuaries, a sea-level rise of 30 centimeters (1 foot) per century would have flooded large areas of land over a period of a few years.[17]

In any event, the growth of the ice caps required an enormous transfer of water from the seas to the polar zones of the continents, necessarily by way of the atmosphere. Each million cubic kilometers of ice on the land represents a lowering of sea level by about three meters (nearly 10 ft.), with the water evaporated to the atmosphere and transferred to clouds, falling as snow and accumulating over the centuries. In addition, in the polar oceans, liquid water freezes directly, the floating ice pack growing every winter, shrinking the following spring and summer. In the Southern Ocean, the volume of floating ice increases from 4 million cu. km, at summer's end in February, to 20 million in September. The ice pack must have extended much farther at LGM, its limit only 1,000 km (600 mi.) from today's temperate zone. Advancing and receding ice are the mark of the climate's instability during the Pleistocene epoch beginning over two million years ago. If we consider only the last two million years, our present interglacial interval, beginning about 10,000 years ago, is the exception rather than the rule. For most of the Pleistocene, ice covered much more area than today. Indeed, from time to time, milder conditions such as today's have prevailed. However, throughout these "interglacial" intervals, Greenland and Antarctica have remained covered by ice, today about 33 million cubic kilometers' worth. The warm interglacial intervals always eventually come to an end: the preceding interglacial—the Eemian—being about 125,000 years ago. The general pattern has repeated itself over the last two million years, and considerable detail is known about the last 420,000. Each interglacial interval ends with the onset of progressive cooling of climate, with successive advances and retreats of the ice at intervals of 21,000 and 40,000 years, plus some quite sudden changes between very cold and very very cold, and overall more advances than retreats. This cooling of the Earth proceeds with fits and starts along a sawtooth descent to a paroxysm of ice, the Last Glacial Maximum of 18,000 B.P. being only the most recent example, not the last. There follows (followed) a radical warming and reduction (but not complete elimination) of ice cover, leading to the interglacial period. It is a sort of wishful thinking to call today's interglacial the "Holocene," as if one could be sure that the Pleistocene has finished. Unless intensification of the greenhouse effect completely upsets things, the long cycle of cooling will almost certainly resume sometime in the next several thousand years, and ice will return to New York and New England, Canada, Scotland, and Scandinavia.

Global ice volume varies with the same periods as the astronomical parameters. The computed peaks of summer insolation at 65°N agree almost perfectly in both timing and strength with the peaks of temperature or of ice volume. The joint Danish-U.S. and, more recently, European expeditions to Greenland, and the American and Franco-Soviet (now Franco-Russian) teams in Antarctica have extracted ice cores corresponding to as much as 420,000 years of climate, and the long climate cycle of 100,000 years is clearly marked in the cores (fig. 5.1). Apart from the ice core record, information on the past is contained in the sediments of the sea floor, rich in fossil shells of different types of plankton, diatoms, and foraminifera. These shells contain silicates and carbonates, thus oxygen. Sediment cores go back tens of millions of years, and so contain a much longer record than the ice. Measurements of the ratios of oxygen-18 to oxygen-16 yield information on the temperatures and ice volumes at these remote epochs. Periodicities of 100,000 and 400,000 years, corresponding to the Milankovitch cycles of the Earth's orbital eccentricity, appear clearly. The strength of these cycles is puzzling because it is not clear why the small changes in orbital eccentricity, which only weakly affect insolation, should have so much influence. Research and debate continue.

At any rate, the sea-sediment data going back 60 million years (60 Myr B.P.), almost as far back as the Chicxulub catastrophe, give a picture of a quite different Earth, nearly free of ice most of the time, a hothouse rather than an icehouse. If continental drift can be neglected, sea level must have been higher than today, perhaps by 70 meters (over 200 ft.). It appears that ice began to appear on Antarctica about 30 Myr B.P., but that the ice cap became a stable feature only about five million years ago; the Arctic Ocean remained ice-free until 3 Myr B.P. For the last two million years, the 400,000- and 100,000-year cycles govern the overall ice cover of icehouse Earth. The shorter 40,000- and 21,000-year periods, corresponding respectively to the obliquity of the ecliptic and to precession, appear clearly in the fluctuations of ice volume (the "sawteeth" of the curve) superposed on the longer cycles (fig. 5.1). The Milankovitch rhythms set the tempo of the climate variations, no doubt about that, but much research is still needed to understand

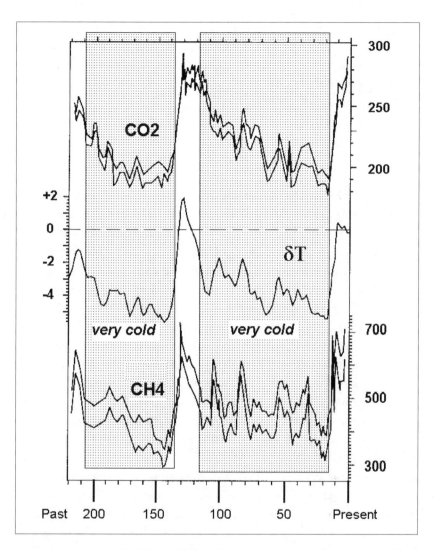

FIGURE 5.1 Ice ages. Glacial-interglacial oscillations of the last 220,000 years (out of the 420,000 decoded by Petit et al. (1999), as determined from the Vostok (Antarctica) ice core. The middle curve (δT) shows the temperature difference (scale on left in °C) relative to twentieth-century climate. Cold or very cold glacial periods (shaded) reigned most of the time; warm interglacial intervals (unshaded) constitute only a small fraction of the whole 220,000-year period. Carbon dioxide abundance (top curves, scale on right in ppm) and methane (CH_4) abundance (bottom curves, scale in parts per billion or ppb) varied practically in unison with temperature over the entire period. In the last part of the twentieth century, both the CO_2 and the CH_4 curves have risen off the scale of this graph, with CO_2 now above 370 ppm and CH_4 above 1,500 ppb, something not seen on Earth for millions of years.

the detailed mechanisms that determine the strength of these changes. In particular, it needs to be shown how some of the astronomical variations, affecting insolation on opposite hemispheres in opposite ways, can lead to nearly simultaneous climate cooling (or warming) in the Northern and Southern Hemispheres.

ICE IN THE TROPICS

When the ice sheets were advancing from Labrador and Scandinavia, what was happening in the Tropics? In the international CLIMAP program (Climate Long-Range Investigation, Mapping, and Prediction), carried out between 1971 and 1976, geologists, oceanographers, and other scientists set about putting together the pieces of this puzzle, constructing the maps corresponding to the Last Glacial Maximum, region by region, season by season (fig. 5.2). These maps show the changes in coastlines for the 120-meter lower sea level, with the English Channel and the Bering Strait above water and with land bridges probably also linking Japan, Borneo, Sumatra, and Java to the Asian continent, and New Guinea to Australia. They also show that areas of cold of both land and ocean expanded simultaneously from the polar zones in the direction of the middle latitudes. Moreover, the reconstructions suggest that at LGM, the atmosphere was much dustier than today, not a surprise considering that the areas of the great deserts of the world, especially the Sahara, increased at the same time that the ice sheets advanced. However, the CLIMAP reconstructions indicate hardly any cooling in the tropical zone, a puzzling result. It is hard to understand the fact that cooling took place simultaneously at higher latitudes both north and south; without some sort of link necessarily acting by way of the Tropics, the astronomical factors, which for the most part promote the growth of ice sheets in the north, would be expected to reduce ice area in the south. Careful and detailed analysis of ice cores extracted from Antarctica and from Greenland show nearly perfect simultaneity of climate fluctuations around North and South Poles, over most of the long 100,000-year climate cycle.

New results may resolve this paradox. In contradiction to the CLIMAP results, new evidence indicates that ice-age cooling did in fact extend to the Tropics. For one thing, it was already known that the snow line on tropical mountain slopes was about 900 meters (3,000 ft.) lower. More recently, the

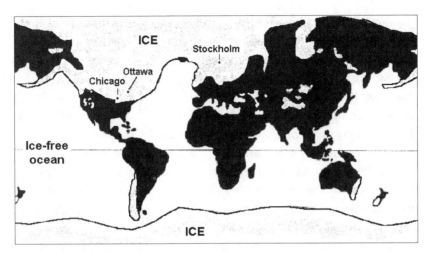

FIGURE 5.2 The world of the Last Glacial Maximum, 18,000 years ago. Thick ice covered the sites of several major present-day cities. Lower sea level left Indonesia connected to southeast Asia, England to France, and Alaska to Siberia.

Ohio State University team of Lonnie Thompson has brought back some solid data from the Andes. They carried several tons of drilling equipment up to a glacier at 6,000 meters (20,000 ft.) at Huascarán in Peru, where they successfully extracted ice cores representing 20,000 years of snow accumulation, including quite specifically the Last Glacial Maximum. They got the ice cores back to the laboratory without melting, and their analyses show that 18,000 years ago, temperatures at the glacier were several degrees lower than today. Analyses of the strontium-calcium ratio in ancient coral skeletons also imply lower paleo-temperatures in the tropical seas, both at Vanuatu in the southwest Pacific and at Barbados in the Caribbean. The heart of the Amazon basin also was cooler 18,000 years ago, judging from ancient pollens trapped in the sediment at the bottom of Lake Pata. With the finding that ice-age cooling did affect the Tropics, it is possible to work out global-scale mechanisms that transmit to the Southern Hemisphere the Milankovitch variations acting on the north.

What sort of mechanisms? The Milankovitch variations directly affect only the seasonal cycle and the pole-to-equator distribution of the incoming solar radiation. One can understand stronger climate response in the Northern Hemisphere (because of its greater land area) than in the

Southern. In contrast, variations of atmospheric gases contributing to the greenhouse effect influence climate over the whole Earth. Analyses of bubbles of air trapped in the ice thousands and even hundreds of thousands of years ago have shown that carbon dioxide and methane were far less abundant during the ice ages (in particular 18,000 years ago) than during the warmer interglacial intervals (cf. fig. 5.1), even before human activities began to increase them 250 years ago.[18] No doubt, the rarefaction of CO_2 and methane during cold periods was a natural phenomenon; cavemen had no impact on the global atmosphere. Cooling almost certainly led to a drying out of the atmosphere, and this decrease in atmospheric water vapor, a powerful greenhouse gas, must have weakened the greenhouse effect and further reinforced cooling over the entire globe. Ice cores bear witness to simultaneous fluctuations of climate and atmospheric composition around both poles, as well as in the Andes and in the New Zealand Alps. Wind strength, as determined from the size of dust grains in the loess deposits of China, and from dust in ice cores, also seems to have increased every time climate cooled. The same atmosphere bathes Greenland, Europe, North America, the Tropics, and Antarctica. The west winds of the middle latitudes connect the North Atlantic, Eurasia, China, and North America. A change in the workings of the water cycle in one region makes itself felt, by way of the winds, on the other side of the world. Atmospheric humidity can change quite quickly in response to a perturbation, and any change in the amount of water vapor of the atmosphere is a change in the strength of the greenhouse effect.[19]

PART II

Water in Today's World

CHAPTER 6

WATER AND ENERGY CYCLES

ICE BUDGETS

On our planet, today both warm and icy, the oceans contain some 1.37 billion cubic kilometers (324 million cu. mi.) of salty liquid water, over 96 percent of the total amount of Earth's water (see table 6.1).[1] Much of the remaining 3.7 percent—that's still 29 million cu. km—lies frozen in the ice caps of Antarctica and Greenland, floating in the ice pack, icebergs, and scattered sea ice of the polar oceans or inching downward in mountain glaciers. The shares of frozen and liquid water change with time, and when the last ice age came to an end between 15,000 and 10,000 years ago, more than 40 million cu. km of ice melted. Most of that water finished in the oceans, and sea level rose by about 120 meters, on average nearly an inch a year for some fifty to sixty centuries. Finally, after perhaps a small overshoot several thousand years ago, the seas reached their present level defining the modern map, with for example the Channel separating England from France, the Strait of Malacca separating Malaysia and Singapore from Sumatra and the rest of Indonesia. Sea level would rise still another 70 to 80 meters (more than 200 ft.) if all the remaining ice on Earth were to melt, if our planet were to be completely transformed from icehouse to greenhouse; and such a sea-level rise would flood huge areas, including many of today's densely populated coastal plains, river deltas, and valleys. During the Tertiary period,

TABLE 6.1 EARTH'S WATER

Estimates given in column 2 are in millions of cubic kilometers (MCkm), where a cubic kilometer is a billion cubic meters. LGM stands for the Last Glacial Maximum, 18,000 years ago. Equivalent depths (column 4) are obtained by dividing the volumes of water by the areas concerned. Rivers, streams, and the biosphere contain minute percentages of the total water, given in column 3 in parts per million (1 ppm = 0.0001%). Estimates of the amount of water underground are very uncertain. The entries correspond to layers from the surface down to 4,000 meters depth. Shallow fresh groundwater, from the surface down to 750 meters depth, amounts to about 4 MCkm, and has shorter residence time. Estimates for fresh and saline water in layers more than 4,000 meters deep, not included in the table, range from 50 to 320 MCkm.

Type of Water	Volume (MCkm)	Share of Total Water (% or ppm)	Equivalent Thickness or Depth (m)	Typical Renewal or Residence Time (years)
Total	1,422	100	2,800	4 Billion
Salt water (oceans, inland seas, salt lakes)	1,370	96.3	3,750	3,700
Of which, water from ice melt since LGM	40	2.8	120	
Present-day "permanent" snow/ice	29	2.0	2,000	From 30 years to 600,000
If it all melts	—	—	80	
Present-day permafrost	0.300	0.21	15	8,000
Soil moisture	0.066	0.005	0.8	1 month
Underground freshwater	10	0.7	65	From 50 years to five million
Underground saltwater	13	0.9	85	Very long
Freshwater lakes	0.12	0.008	75	10 to 1,000
Rivers and streams	0.002	1.4 ppm	13 mm !	18 days
Biosphere	0.001	0.7 ppm	2 mm !	Hours
Atmospheric water vapor and cloud water/ice	0.013	91 ppm	26 mm !	8 to 10 days

"only" a few tens of millions of years ago, such shallow seas covered a significant fraction of today's dry land. No such extensive flooding is expected for the next few ten thousand years at least. As for the 21st century, even with strong greenhouse warming, sea-level rise by the year 2100 will likely be about two feet (less than a meter in any event), mostly due to thermal expansion of water in the sea's surface layers, and not to melting ice.

WATER BUDGETS: CLOUDS, RAIN, AND SNOW

Setting the stage for our present interglacial interlude, the grand ice debacle literally put into circulation vast amounts of fresh water, giving a last rinse to the lands sculpted by ice streams and glaciers. Some of that fresh water percolated underground, constituting a reserve of "fossil" water exploited, indeed *over*exploited, in many areas of the world. This nonrenewable resource is often treated today as though it were inexhaustible. Apart from such infiltration, most of the fresh water from yesteryear's long-gone glaciers flowed to the seas and entered the water cycle with its atmospheric phases. Meltwater from mountain glaciers feeds many rivers—for example, the Rhone River in Switzerland and France—and that water can be tapped for supplying cities, for irrigation, and for cooling nuclear or fossil-fuel-burning power plants. But the Rhone glacier would have disappeared long ago were it not regularly renewed by snowfalls continuing (and perhaps intensifying) after the retreat of ice from the Valais region of Switzerland. Life on the land depends on water from the sky.

In the present-day climate, about 40 inches (1,000 mm) of rain fall—on average—on the Earth, somewhat more over the oceans, less (about 30 in. on average) over land. But these are averages, and the range of rainfall is enormous. Decades can go by in the desert without a single drop of water, but in Assam, on the slopes of the Himalayas, over 11 meters (430 in.) of rain fall in the year, most in the two months of the summer monsoon. In South America, annual average rainfall exceeds 60 inches, and over the Amazon basin it approaches 80 inches. In Hilo, on the windward side of the big island of Hawaii, annual rainfall tops 120 inches. But in Antarctica, average annual snowfall amounts to less than 7 inches of water.

A meter of rainfall over the entire globe—the annual average value—constitutes half a million cubic kilometers of water, enough to renew the oceans completely in a little less than 3,000 years. Water from heaven—from the skies of meteorology, not of astronomy—total rainfall over the continents (112,000 cu. km per year; see fig. 6.1) amounts to almost three times total river runoff. But at any given moment, the atmosphere contains only about 13,000 cu. km of water, most of it in fact as water vapor, i.e., H_2O molecules as a gas mixed with the rest of the air. The atmosphere loses about 1,000 mm of water in the course of the year (close to 3 mm on average per day), mostly in the form of rain. Such precipitation depends on the continual resupply of the atmosphere by way of evaporation of water from land and sea surfaces. The ratio of the two numbers (1,000 to 26) shows that the total atmospheric water content must be renewed some forty times over the year; on average, a water molecule spends only nine days in the atmosphere. Water passes rapidly and frequently through the atmospheric part of the water cycle, but this passage is essential to bring water to the land. By con-

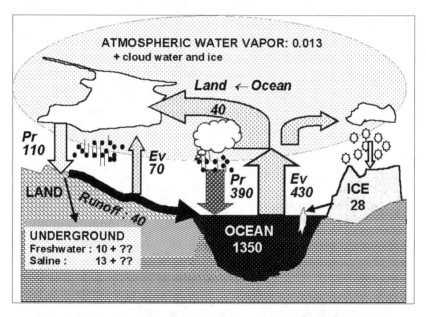

FIGURE 6.1 Earth's water, the water cycle, and elements of the water budget. Water stocks (see also table 6.1) in the different reservoirs are given in millions of cubic kilometers of liquid water equivalent. Fluxes of water (in *italics*) are given in thousands of cubic kilometers per year (1 cubic km per year is equivalent to 31.7 metric tons per second). The fluxes include evaporation, precipitation, runoff, and infiltration (not quantified here). The excess of precipitation relative to evaporation over the oceans makes possible the atmospheric transport of water from the oceans to the continents.

trast, a drop of water in the ocean has a good chance of staying there for centuries, or even for thousands of years if dragged down to the deep sea.

Clouds cover more than 60 percent of the surface of the globe, but at any given moment, compared with the amount of water vapor in the atmosphere, the total amount of water they contain in the form of liquid water droplets and/or solid ice crystals is minute. Nevertheless, practically all precipitation comes from clouds, mostly from a small part of the 60 percent cloud cover. Renewal of cloud water proceeds at an even faster pace than renewal of atmospheric water as a whole, with clouds continually forming, growing, and dissipating. Over the subtropical oceans, off the western coasts of the continents—Africa (Morocco, Angola), North and South America (California, Peru), and Australia—low clouds (stratus and stratocumulus) develop every night, disappearing in the course of the day, often without a single drop of rain. Atmospheric water vapor condenses, forming water droplets at the cloud's bottom, but other droplets are continually evaporating at the cloud's top. At night, condensation dominates and the clouds thicken, but by day, with the warming effect of the Sun's rays, evaporation takes over. In France, from the Atlantic coast inland as far as Paris, morning cloud cover often dissipates in the course of the day. In the summer, further away from the ocean over midlatitude continental areas, over much of North America, and in Europe from Germany to Russia, fair-weather cumulus clouds develop in midday and afternoon as a result of land evaporation, and tend to dissipate (i.e., the water droplets evaporate) in the evening, again without bestowing any rain. These relatively thin clouds continually exchange water with the atmosphere but never contain more than a minuscule amount (less than a tenth of a millimeter water equivalent) at any instant. To produce rain, 3 millimeters per day on average, sometimes as much as a few tens or even hundreds of millimeters in a few hours, water must be supplied to form clouds carrying much larger quantities of water in liquid or solid form.

WATER BUDGETS: EVAPORATION, CONDENSATION, AND LATENT HEAT

Maintenance of atmospheric moisture requires evaporation of about 3 millimeters of water per day on average over the globe. Obviously, such evaporation can only proceed where water is available—from the surfaces of lakes, rivers, and sea, from wetlands, vegetation, and, to a much smaller extent, from snow and ice. However, to extract an H_2O molecule from liquid water so as to make it a free molecule in air requires energy, to extract a molecule from the crystalline structure of ice or snow still more energy— energy generally in the form of heat: to transform a gram of liquid water into water vapor requires 540 calories (2,500 J or joules); and to vaporize (sublimate) a gram of snow or ice, 620 calories are needed. Evaporation of water from a surface uses up heat that would otherwise warm that surface, so in effect it keeps the surface cool. Thus, evaporation of our perspiration helps keep us warm-blooded animals from overheating when the temperature of the air rises. However, this only works if the air accepts the additional water vapor; when relative humidity gets close to 100 percent (in other words, when the air is nearly *saturated* with water vapor), evaporation becomes very weak, or rather, there is nearly as much condensation of water vapor on our skin as there is evaporation of perspiration from our skin. As a result, without air conditioning, sleep comes hard during the hot humid nights of the Tropics or of New York summers, when relative humidity sticks close to 100 percent and temperatures remain above 30°C (86°F). In general, evaporation from a surface is highest when that surface is moist and hot, the air dry, and when movement of the air (wind, convection, turbulence) removes water vapor from the immediate vicinity of the surface, mixing it with drier air farther off.

The H_2O molecules evaporated from a moist surface carry with them the "latent heat of evaporation," the energy used to evaporate them. It is latent because the energy thus removed from the surface cannot disappear. When and at the place where the water vapor condenses, whether it be a cold surface, such as my eyeglasses in winter, or a cloud in the sky, that latent heat reappears as warming. Corresponding to the evaporation of 3 millimeters of water (on average) every day from the surface of the Earth, a flux of latent heat is removed from the surface and transferred to the atmosphere. In terms of energy, the flux amounts to 90 watts per square meter on average, a total of 45 million billion watts for the entire Earth, significant compared to other energy fluxes in the Earth's atmosphere and between the Earth and space (see fig. 6.2). Apart from dew or frost forming on cold surfaces mostly at night, apart from fog or mist, the latent heat is generally released as part of condensation of water vapor in clouds, forming where moist air has ascended to higher cooler layers of the atmosphere.

In principle, condensation occurs precisely at the level where the air temperature goes below the dew point, where the air becomes saturated and relative humidity reaches 100 percent. In reality, condensation usually occurs under "supersaturated" conditions, with relative humidity above 100 percent, i.e., at temperatures somewhat below the dew point. In any case, the latent heat released heats up the air, i.e., it intensifies the agitation of the molecules making up the air, made up mostly (99%) of nitrogen (N_2) and oxygen (O_2) molecules. Molecules moving at higher speeds effectively take up more space, i.e., the heated air expands and becomes less dense, and by buoyancy it tends to rise as does a *montgolfière* or hot-air balloon.

Evaporation and condensation recycle the atmosphere's water forty times a year. Throughout the year, evaporation removes heat from the Earth's surface, but on average, it does not freeze. Condensation deposits heat in the atmosphere, but on average, it does not indefinitely warm.

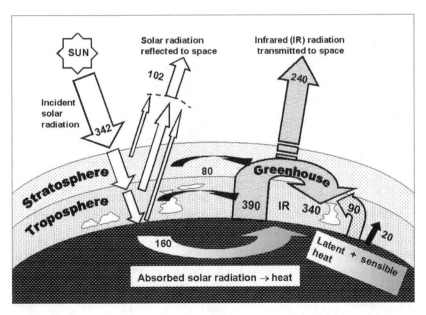

FIGURE 6.2 The energy budget of the Earth-atmosphere system, with global annual mean energy fluxes given in watts per square meter. The horizontal (left to right) arrows correspond to conversion of solar radiation into heat in the stratosphere, troposphere, and at the Earth's surface. Evaporation of water at the surface and its condensation in clouds gives the latent heat flux of 90 W/m² from surface to atmosphere.

Global warming or, on the contrary, ice age onset, generally require imbalance over decades. The water cycle accounts for the fact that the seas do not overflow, neither do they empty out, and that overall, condensation in and precipitation from the atmosphere make up the losses by evaporation, not locally but globally. The water cycle operates practically entirely within what may be called the "Earth system," with neither significant inputs of water from outer space, nor significant losses. But the water cycle requires energy to make it go round, and to compensate the latent heat loss by the Earth's surface. Energy is needed to keep the Earth warm, in particular its oceans. Closely coupled to the water cycle's turning with a constant stock of water, Earth's energy cycle involves a roughly constant *flow* of energy *through* the system, coming from the Sun and ultimately escaping to outer space.

ENERGY BUDGETS AND THE GREENHOUSE EFFECT

Common sense tells us that the Sun keeps the world warm; and on this point, common sense is right. Without the Sun's radiation, the world would not stay warm enough to keep water from freezing. The flux of heat from the interior of the Earth, only 0.09 watts per square meter on average, is nowhere near enough. The Sun sends our planet an average energy flux of 342 watts per square meter (W/m^2), quite enough to supply the 90 W/m^2 needed to evaporate almost 3 mm of water per day. However, to keep the water cycle going in this way, the Sun's radiation must reach the surface of the planet, without too much being lost in the atmosphere. Solar radiation consists mostly of visible light (and some near ultraviolet radiation) at wavelengths going from 0.3 to 0.8 micrometers (μm), and a fair amount of near infrared radiation (wavelengths from 0.8 to 4 μm), invisible but warming. In the absence of clouds, much of the solar radiation passes freely through the atmosphere, apart from some bands in the near infrared, absorbed by water vapor, and ultraviolet radiation near 0.3 μm absorbed by ozone. Scattering by the nitrogen and oxygen molecules that make up 99 percent of air does affect the shorter visible wavelengths, giving us the blue sky. As for clouds, they reflect and scatter back to space a significant fraction of the incident solar radiation, while at the same time allowing transmission of some sunlight down to the surface. Even for a completely cloud-covered sky, it rarely gets really dark when the Sun is up. Considering the effects of clouds as well as scattering and absorption by the atmosphere, not quite half of the incident solar radiation, about 150 W/m² (the exact figure being uncertain and a matter of lively debate among specialists) reaches the surface. Is that enough to account for the average surface loss by evaporation of 90 W/m² of latent heat flux?

At the very least, the surface temperature has to be kept above the freezing point of water. Indeed, the average temperature at the Earth's surface is +15°C (59°F), or 288 K (degrees Kelvin on the absolute scale). Now any body warmer than absolute zero (0 K, −273°C, −459°F) emits radiation, and the warmer the body, the shorter the wavelengths of emission. At 500 K or 227°C (441°F), a poker heated in the fire is visibly red-hot, although in fact emitting mostly at longer wavelengths (0.8 to 6 μm) in the near infrared. The white-hot Sun, with temperatures ranging from 4,000 to 8,000 K in its photospheric layers, radiates strongly between 0.3 and 0.8 μm. We humans, warm-blooded animals with normal body temperature 98.6°F (37°C, 310 K), radiate in the medium infrared at wavelengths ranging from 3 to 30 μm. The Earth's surface, not quite so warm at an average temperature of 288 K (+15°C), emits about 390 W/m² of infrared radiation upward, mostly at wavelengths from 4 to 40 μm.

How can the energy budget be balanced? It would seem (fig. 6.2) that the Sun supplies only 150 W/m² to the surface of the Earth, while the surface loses at least 480 W/m² by evaporation and upward radiation. What keeps the world warm? The answer is in the atmosphere, in that perfectly natural phenomenon, the greenhouse effect. The surface receives some 330 W/m² of infrared radiation coming downward from the atmosphere, absorbed and reemitted by the polyatomic molecules (molecules made of three or more atoms) in the more or less humid air. The most important of the "greenhouse gases," water vapor (H_2O), absorbs and emits at infrared wavelengths between 5 and 7 μm and beyond 16 μm; carbon dioxide (CO_2) between 13 and 17 μm; and ozone (O_3) at 9.6 μm. Air is relatively transparent at "short" wavelengths (but not too short, certainly not in the ultraviolet), but it blocks a large part of the "long-wave" thermal infrared radiation emitted by the Earth. Air consists mostly of nitrogen (78%) and oxygen (21%), but the structure of these N_2 and O_2 molecules is much too simple for them to interact effectively with infrared radiation, and they contribute virtually nothing to the greenhouse effect.

Atmospheric water vapor plays the major role in the natural green-house effect that keeps the Earth's surface warm enough for water to stay liquid. Absorbing a large part of the infrared radiation emitted upward by the surface, the atmosphere sends enough energy back down, again in infrared form, for the surface to continue to lose energy by radiation and evaporation. Surface evaporation in turn supplies water vapor to the atmosphere. Balancing the surface energy budget, the Earth's surface loses on average another 20 W/m² in the form of "sensible" heat. What happens is that because the surface and the air immediately in contact with it are a bit warmer here, a bit cooler there, convection can set in, removing heat from the surface even when conditions are perfectly dry. Where the air is warmer and so less dense than average, it rises, while cooler less dense air descends to take its place. Even in the absence of wind, free convection can cool the surface, and when the wind blows, especially when it blows cold on a warm surface, convection and surface cooling are further strengthened ("wind chill"). When temperatures fall, we humans lose relatively little in the way of latent heat because our body mechanisms essentially turn off perspiration so that hardly any water evaporates from our skin. Of course, our breath contains water vapor, and when I'm out in bitter cold, the vapor freezes to form icicles on my moustache, and condenses as fog on my glasses. That can be a nuisance, but loss of sensible heat is a real danger when a strong cold wind blows away the layer of warmer air next to exposed skin. And when temperatures drop below −30°C (−22°F), as frequently occurs in winter in Alaska, Canada, or northern Russia, the air can sting your face, even without wind!

Energy budgets must balance, not only for the surface but also for the atmosphere. The atmosphere absorbs and so transforms into heat a small but nonnegligible fraction of the energy of solar radiation, a few tens of watts per square meter, the exact amount that gets through the atmosphere being a matter of debate. The Earth's atmosphere emits 240 W/m² of infrared radiation to space,[2] and this rate of energy loss almost exactly compensates the rate at which solar energy is absorbed and converted into heat by the system (Earth with its oceans, ice, and atmosphere). Many specialists put a number on the planetary greenhouse effect by taking the difference between the flux of thermal infrared radiation emitted upward by the warmed surface (390 W/m²) and the flux (240 W/m²) of absorbed solar radiation that finally returns to space in the infrared; that gives a greenhouse effect of 150 W/m^2. The atmosphere also radiates about 330 W/m^2 of infrared downward to the surface. The natural greenhouse effect constitutes a recycling of absorbed solar energy in the form of infrared radiation downward from the atmosphere to the surface so that the surface can radiate upward more than the actual absorbed solar energy flux. The recycling results from the opacity of the atmosphere at the infrared wavelengths of the Earth's thermal radiation. Without such recycling, the surface temperature corresponding to the absorbed solar energy would be a frigid −18°C (0°F), known to specialists as the "effective" temperature. Another common way of quantifying the greenhouse effect is to take the difference (33°C, 59°F) between the actual surface temperature and the effective temperature. It should, however, be noted that an Earth without an atmosphere providing a greenhouse effect would be an Earth without clouds, ice, or vegetation. Such an Earth would be as dark as our Moon, absorbing about 93 percent of the incident sunlight, reflecting only 7 percent. Like our Moon, its effective and average surface temperature would be close to 0°C (32°F, 273 K), still pretty cold.

In the Earth system, the water cycle operates in a closed circuit, with a practically constant stock (fig. 6.1). But the energy cycle (fig. 6.2) involves inflow of energy from the Sun, outflow to space. To maintain a living watery Earth despite the continual heat loss to space by way of thermal radiation, continual inflow of energy is essential, and the Sun supplies it. Without the Sun, Earth would be a cold frozen globe, at best with a few volcanic hotspots. But an Earth without water would not be more attractive, a dry Moon-like planet warming to torrid temperatures by day, cooling well below zero by night, with only the same volcanic hotspots to liven things up a bit.

LATITUDES AND SEASONS

The Sun warms the Earth with its radiation, furnishing a flux of energy worth 342 watts per square meter on average. This average energy flux is just the solar "constant" of 1,368 W/m^2 divided by four, and this division by four corresponds to the ratio of the Earth's surface area $4\pi R^2$ to its cross-section πR^2, R being the Earth's radius. In reality, of course, the energy flux reaching the "top" of the atmosphere at any place on Earth varies between zero at night (and at any moment, night prevails over half of the globe), and a maximum value at local noon. That maximum value is the solar constant adjusted for the Sun-Earth distance at those locations where the Sun passes through the zenith. The absolute maximum occurs on January 9, when planet Earth goes through perihelion, i.e., through the point of its elliptical orbit closest to the Sun. On that date, 1,410 W/m^2 reach the top of the atmosphere at local noon along the parallel at latitude 23°27' South. However, the Sun only passes through the zenith in the zone between the Tropics, i.e., between latitudes 23°27' North and South, shifting from north to south from June to December and back again. That includes such places as Honolulu, Havana, Acapulco, San Juan (Puerto Rico), and Darwin (Australia). Outside the tropical zone, whether it be in Paris or generally in Europe, in North America north of the Rio Grande, in New Zealand, or in Melbourne (Australia), the Sun never goes through the zenith. Over the whole year, average solar radiant energy flux reaching the top of the atmosphere ranges from 140 W/m^2 at the North Pole (150 at the South Pole) to more than 400 W/m^2 at the equator. At the time of the winter solstice (June 21 in the South, December 21 in the North), no solar radiation at all reaches the zone of polar winter night beyond the Antarctic and Arctic Circles at latitudes 66°33' South and North, respectively. In the winter hemisphere, average insolation over a 24-hour period ranges from zero to a maximum close to 400 W/m^2 near the equator, and the Sun does not reach the zenith. In the summer hemisphere, on the other hand, the Sun passes higher in the sky and stays longer above the horizon. Beyond the polar circle, it's the land (or ice) of the midnight Sun. On the day of summer solstice, average insolation exceeds 400 W/m^2 over the entire summer hemisphere, and on December 21, at the South Pole, it reaches 560 W/m^2. Of course, only part of the incident radiation penetrates the atmosphere and is absorbed and converted into heat. Near the poles, ice, snow, and clouds reflect much of the incoming solar radiation back to space, and that magnifies the contrasts between the poles and the equator with respect to the solar energy actually available for keeping the world warm. Near the poles, average net solar energy input over the year is only 40 W/m^2; while near the equator, eight times as much solar energy flux—over 320 W/m^2—is converted into heat.

On average, the atmosphere contains the equivalent of a layer of 26 mm (a little over an inch) of water, but that water content, mostly in gaseous form (i.e., as water vapor), depends on continuous resupply by evaporation, ultimately depending on energy supplied by the Sun. The enormous variation of solar energy flux between winter and summer entails strong variations in atmospheric water. In the Northern Hemisphere, the average amount of water in the atmosphere water goes from 20 mm equivalent in winter to nearly 35 mm in the summer; and in the Southern Hemisphere, it goes from 21 mm in July to 28 mm in January. That may seem not very much, but for the Northern Hemisphere it amounts to a relative variation of 75 percent; in different terms, the atmosphere of the Northern Hemisphere contains 15 kilograms water per square meter (about 3 pounds per square foot) more in summer than in winter, altogether some 3,750 billion metric tons more for the entire hemisphere. By contrast, relative variation is only 33 percent in the Southern Hemisphere. The two variations do not compensate exactly, and for the entire globe the atmosphere contains more water in July and August than it does in December and January. The cause of the North-South difference? It's not the Earth's elliptical orbit, which brings the Earth closer to the Sun in January, making more radiant energy flux available then for heating. What accounts for the greater humidification of the atmosphere during northern summer, especially considering the greater area covered by ocean in the Southern Hemisphere? This apparent paradox can be resolved, considering that atmospheric layers near land surfaces undergo much stronger heating than over the sea. As a result, the larger land area in the North acts to warm the air more than in the South, and heating of the air determines how much water it can hold. Oceans of the North lose water to the atmosphere between June and September, oceans of the South between December and February. For the world as a whole, the oceans contain a little less water in July than in January, about 3 mm (⅛ in.), but what's that considering how deep is the sea?

Between evaporation and precipitation, atmospheric water completely renews itself in a little over a week on average. The seesaw of humidity between north and south, driven by the astronomical cycle of insolation, depends both on the strong seasonal temperature changes over the continents and on water exchange between the oceans and the atmosphere. However, changes in air temperature necessarily also entail changes in density: at fixed pressure, higher temperatures imply lower density. Between April and June, air above the Northern Hemisphere's continents gets progressively warmer and less dense while taking up more humidity; and dry air tends to migrate to the South. The effect is tiny, corresponding to a flow of less than 1.5 mm per second (about 18 ft. per hour, or 2 mi. a month); it reverses between August and October, and does not extend beyond tropical and subtropical latitudes. Even so, the total atmospheric mass above a hemisphere is slightly greater in winter than in summer. For the entire globe, the effect of the Southern Hemisphere's winter is stronger, and the atmosphere is slightly more massive between June and August than from November to March. The total amount of dry air, nearly entirely nitrogen, oxygen, and argon, stays constant throughout the year, but because of the humidity variations, dry air must flow back and forth between north and south and between winter and summer. Because these atmospheric flows operate on very large scales, they affect the rotation of the Earth. The Earth's spin slows when the atmosphere piles up above the equator and Tropics because that increases the moment of inertia of the Earth-atmosphere system. It speeds up, on the other hand, when a slight excess of atmospheric mass piles up around Antarctica during southern winter. Tiny as they may be, the corresponding fluctuations of the Earth's rotation are accurately measured by astronomers.

Land and Sea

What we call dry land, continually drained by the rivers that flow to the sea, would indeed dry out completely were it not for the return to the continents, by way of the atmosphere, of water from the oceans. On average, about 120 cm (47 in.) of water evaporate from the sea's surface in a year, not completely compensated by the 107 cm or 42 inches of annual rainfall there. Most of the water evaporated from the sea's surface does indeed return by a nearly direct route, and quite quickly, in a matter of days. Still, considering that oceans cover 70 percent of the Earth's surface, the difference of 12 cm (5 in.) of water per year accounts for the 40,000 cu. km of water that flow in the rivers from land to sea. However, precipitation (rain and snow) over land, on average 75 cm (30 in.) of water per year, adds up to a volume of 112,000 cu. km of water, nearly three times as much (fig. 6.1). This comes about because water falling on the continents has nearly two chances out of three of being reevaporated before reaching the ocean.

The atmospheric recycling of rainfall plays a particularly important role far from the sea in the interiors of the continents, and it depends to a large extent on vegetation. Plants use the water they pump from the ground in two different ways. By the process of photosynthesis, they transform some of this water into their own organic matter. However, they return larger amounts of water directly to the atmosphere through their stomata, microscopic openings on their leaf surfaces that serve also as points of entry and exit for carbon dioxide and oxygen in the processes of photosynthesis and respiration. The process of evaporative evacuation of water, analogous to our perspiration and called *evapotranspiration*, acts to maintain the plant's temperature within tolerable limits. In the interior of the Amazonian basin, far from the Atlantic and cut off from water of Pacific Ocean origin by the high barrier of the Andes, the rain that falls consists of water several times recycled through vegetation since its evaporation from the Atlantic Ocean off the coast of Brazil. Each time water evaporates, the slightly heavier H_2O molecules containing heavy isotopes of either hydrogen (deuterium: D or H-2 rather than ordinary H-1) or oxygen (O-18 or O-17 rather than ordinary O-16) tend to get left behind. In the Amazonian interior—for example, at Leticia, where Colombia, Brazil, and Peru meet—rainwater contains a measurably lower proportion of the heavier isotopes, having been several times through evapotranspiration by vegetation since its extraction from the Atlantic. The same heavy water deficit occurs wherever the rain that falls is composed of multiply recycled water without direct supply of water evaporated from the sea. And that raises the question of the risk that deforestation, reducing the chances of recycling water by evapotranspiration, may lead to reduction of rainfall in the continental interiors. The answer is not simple. The analogous inverse reasoning ("Rain follows the plow"), more wishful thinking than wisdom, led many to invest their own perspiration in farming on semiarid lands of the western Great Plains in the United States. The topsoil blew off as dust at the first prolonged severe drought. That led to the dust bowl tragedy of the 1930s described in John Steinbeck's novel *The Grapes of Wrath* (and the film), and in Woody Guthrie's song "So Long, It's Been Good to Know You."

The planetary water budget may appear simple, but on closer inspection, many complications appear. For the globe as a whole, supply of water to the land depends on the slight excess of evaporation relative to precipitation over the seas, multiplied by the recycling of the rainfall over mostly vegetated land. The inspection of individual ocean basin budgets turns up some surprises. For example, the Mediterranean exports water vapor: rainfall there is quite small, but evaporation removes a meter (39 in.) of water a year. Even taking into account the flow from the Nile River and a few others, without water inflow from the Atlantic by way of the Strait of Gibraltar, the high rate of evaporation would be enough to dry out the Mediterranean completely in a few thousand years. And that did indeed happen six million years ago, when tectonic motions closed the opening to the Atlantic. In fact, only the Atlantic and the Indian Oceans export large amounts of water vapor. Although evaporation from the vast surface of the Pacific Ocean is enormous, still more rain falls there: 1,290 mm (over 50 in.) of precipitation per year against 1,200 mm of evaporation. The supply of water to Asia depends in part on the Indian monsoon, in part on some water of Atlantic origin managing to cross Europe without precipitating. No surprise then to find vast expanses of desert and semiarid land in Asia, especially in those areas of Central Asia where high mountain ranges block the penetration of the water-laden air of the Indian monsoon.

CHAPTER 7

WINDS, WAVES, AND CURRENTS

The Tropical Furnace and the Hadley Circulation

Water circulates incessantly between heaven and Earth, or rather between ocean, air, and land. From tropical sea surfaces warmed by abundant sunshine, water evaporates, and the wind carries the water in the gaseous state toward the equator. The warm humid air rises, and as it rises it cools, and once the temperature falls below the dew point, the water vapor condenses. Where the winds converge and strong convection lifts air upward, towers of cumulonimbus clouds rise to altitudes of 15 km (49,000 ft.) or higher, marking what is often called the "meteorological equator." The water comes back down in the form of rain, and the air that rose, now dry, must also descend. Much of the descent occurs locally in the same storm system, in strong downdrafts that pilots try to avoid. Updrafts and downdrafts inside tropical convective clouds can have speeds as high as 10 meters (33 ft.) per second. Overall, however, upward motion dominates in the zone of maximum heating. In this zone shifting north and south of the equator, more or less following the Sun, more air is moving up than down.

Although air rises in the equatorial zone with an average upward motion of only a centimeter (less than half an inch) per second, it must be replaced. Replaced it is by the convergence of the trade winds coming from the tropical latitudes north and south of the equator. Passing over the sunny warm tropical seas, these low-level winds carry abundant moisture and with it latent heat of evaporation. Although they sometimes use the term "meteorological equator," meteorologists prefer to call this the "intertropical convergence zone" or ITCZ. With the moisture-laden trade winds converging on the ITCZ, the only direction the air can take is up, and in another sense the upward motion sucks in the converging winds. The explanation may appear to involve circular reasoning, but atmospheric circulation is what this is about. The first cause is the geographical distribution of sunshine, but beyond that it is futile to try to distinguish causes from effects in this system of interrelated processes.

Following a mass of moist air as it rises above the equator, pressure decreases, for atmospheric pressure corresponds precisely to the weight of the air above. Temperature also decreases with altitude, generally between 5 and 8°C per kilometer (roughly 3 to 4°F per thousand feet), and so the rising air eventually reaches the dew point level at which its moisture can no longer remain entirely in the gaseous state. The water vapor condenses, forming the liquid water droplets or ice crystals that constitute clouds. The condensation releases latent heat, warming the air and further stimulating its ascension, giving rise to the towering cumulonimbus clouds of the ITCZ that climb as high as 18 km (60,000 ft.). Having gone up part or much of this way, most of the condensed moisture comes back down as rain. The remaining dry air, having reached the uppermost layers of the troposphere, must leave the equatorial zone to make way for the rising air masses that follow. At the tropopause, the upper limit of the troposphere, the air flow diverges: some goes north, some south, the flow generally being stronger toward the winter hemisphere. Finally, the air having risen over the equatorial zone descends in the neighborhood of the Tropics, near 23½° north and south latitude, in a slow motion called atmospheric subsidence. As the dry air descends, its temperature rises; its already low relative humidity becomes even lower, practically excluding any chance of condensation or precipitation. Zones of atmospheric subsidence are zones of aridity, and most of the world's great deserts lie there, both north (the Saharan and Arabian Deserts, near the tropic of Cancer) and south (the deserts of Namibia and Australia, and the Atacama Desert of northern Chile, all near the tropic of Capricorn).

With air piling up along the Tropics, surface atmospheric pressures are slightly higher than elsewhere, and from this band of atmospheric subsidence and high surface pressure, the air spills out in divergent motions in the lowest layers. This accounts for the trade winds over the oceans, and the dry *harmattan* wind of the Sahara, completing the loop. On a north-south cross-section of the atmosphere along a meridian, one can sketch the atmospheric motion as two cells circulating north and south of the equator (fig. 7.1), called Hadley cells in honor of the English scientist George Hadley, who first imagined this arrangement to explain the trade winds.[1] To sum up: warm moist air rises in the equatorial zone, losing its moisture in abundant rainfall; the dried-out air at upper atmospheric levels moves away from the equator and descends in the high-pressure zones near the tropics of Cancer and Capricorn; once at low levels, much of that air then returns to the equatorial zone in the trade winds, picking up moisture from the sunny warm seas along the way. Actually, this Hadley circulation is stronger by far in the winter hemisphere around the winter

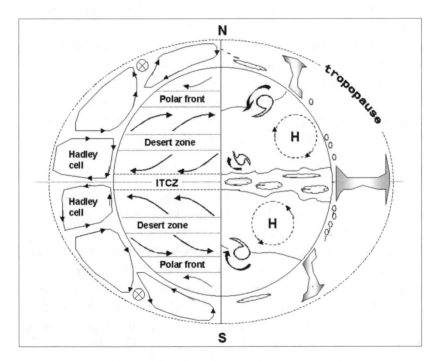

FIGURE 7.1 The general circulation of the atmosphere. On the left, a very schematic representation of near-surface winds averaged over the year, with a cross-section showing how mean meridional (north-south) motions vary with height in the troposphere. Typical positions of the polar jet streams are marked ⊗. On the right, typical cloud cover, also in cross-section. Vertical scale in the cross-section is highly exaggerated; it corresponds to about 10 km near the poles and 18 km near the equator, whereas the Earth's radius is 6,370 km. The "polar front" corresponds to the mean trajectory of midlatitude cyclonic disturbances (the average storm track).

solstice. Nevertheless, both cells appear in the circulation of the atmosphere averaged over all longitudes.

On average, air moves toward the equator near the surface, away from it at upper levels, but not just in a north-south or south-north direction. Near the surface, the air tends to follow the spin of the solid Earth, 40,000 km in twenty-four hours, over 1,000 miles per hour from west to east. As the air masses having reached the tropopause high above the equator move away either north or south, they follow the curve of the Earth and so approach the Earth's axis of rotation. In such a case, the law of conservation of angular momentum requires that they speed up in their motion eastward around the Earth's axis, just as a the spinning of a figure skater speeds up when he brings his arms closer to his body. On the other hand, near the surface where the air masses moving toward the equator necessarily move further away from the Earth's axis, they must slow down in their eastward motion, and relative to the solid Earth they are deviated westward. Thus north of the equator, the southward drift of air gives trade winds blowing from the northeast, while to the south, northward drifting air gives the trades from the southeast. The Coriolis[2] force that deviates all motions to the right in the Northern Hemisphere, to the left in the Southern, must be included when writing the equations of motion of the atmosphere (or artillery shells or anything else) on the spinning Earth.

East Winds, West Winds, and the Gyrations of the Oceans

The Hadley cell does not stretch all the way to the poles, and trade winds don't blow over all the seas. The Earth turns, and with it everything else, oceans as well as atmosphere. The strong solar heating in the latitude band around the equator gives rise to the Hadley cell overturning: air rises from the surface to the tropopause, then moves either north or south before descending and drifting back toward the equator. The Coriolis force due to the Earth's rotation transforms the north-south atmospheric drift into the northeasterly and southeasterly (westward blowing) trade winds near the surface. However, neither the overturning of the Hadley cell nor the trade winds can extend all the way to the poles. Generally, sunshine is stronger near the equator than near the poles, but at any location the Sun spends as much time to the west as to the east. East-west winds must cancel out over the whole planet. Trade winds blow systematically from the east, exerting a frictional force that tends to slow down the solid and liquid Earth's rotation. Somewhere on the globe, winds must blow in the other direction. Indeed they do, tending to come from the west in the zone of middle latitudes, these westerlies compensating the easterlies of the tropical zone, in a sort of global-scale gyration. In reality, things aren't so simple. Especially in the middle latitudes, weather systems take the form of moving smaller-scale eddies, with strong rotating motion around traveling low-pressure centers. New Englanders keep a lookout for the development of what are known as "nor'easters." The winds in the most dangerous sector of the rotating storm tend to be those loaded with water from the Atlantic and blowing from the northeast before the center of the storm itself arrives from the south or southwest; and in the winter, the water may come in the form of heavy snow. Other aspects of atmospheric dynamics involve concentrating and channeling a considerable amount of momentum into narrow fast-moving currents of air—the jet streams—and as a result it can take an hour or an hour and a half less to fly a Boeing or Airbus east from New York to Paris than it does to fly back. Although the jet streams flow in a sense above the weather, they play an important role in the birth and development of atmospheric "waves" and the resulting traveling low-pressure weather systems.

The trade winds, pushing surface waters of the tropical seas to the west, tend to slow down the Earth's spin. Counteracting this, the strong prevailing westerlies and "wild west winds" of the middle latitudes drive ocean surface waters eastward and accelerate the planet's rotation. Sailing south toward Antarctica means confronting the "roaring forties," the "furious fifties," and the "screaming sixties" where no land masses can interrupt the westerly air flow and the eastward drift of the Southern Ocean. The global-scale pattern of prevailing winds, easterlies in the Tropics, westerlies at middle latitudes, drives the pattern of surface currents, circulating in *gyres* both in the Atlantic and the Pacific (fig. 7.2).

In the Tropics, both south and north of the equator, surface ocean waters, driven by the trade winds, drift for the most part to the west. However, unlike the Southern Ocean south of Cape Horn where no land masses block the eastward drift, the tropical seas are limited by continents and divided into the Atlantic, Pacific, and Indian oceans. In the Atlantic, waters of the equatorial current warm up as they cross the Atlantic (running up

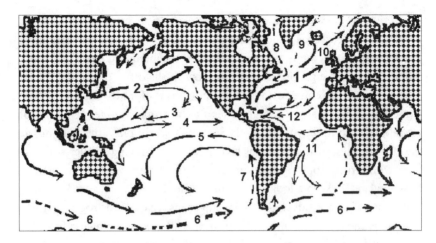

FIGURE 7.2 The main ocean surface currents. (1) Gulf Stream; (2) Kuro-shiyo; (3) North Pacific equatorial current; (4) equatorial countercurrent; (5) south equatorial current; (6) West Wind drift; (7) Peru current; (8) Labrador current; (9) Greenland current; (10) North Atlantic drift; (11) Brazil current; (12) North Atlantic equatorial current.

against the Americas, where they exert a force slowing down the Earth's rotation). The water must go somewhere, either north or south. A northern branch circumvents Cuba and makes a tour of the Gulf of Mexico, exiting to form the Gulf Stream off the southeast coast of the United States. Moving north, the current comes under the combined influence of the Coriolis force and the prevailing westerly winds; and starting at Cape Hatteras and the North Carolina coast, it turns gradually away from America and finally crosses the Atlantic in the direction of Europe. During the entire crossing, Gulf Stream water transfers warmth and humidity to the air above, and they are picked up by the west winds. As a result, Ireland and Brittany are blessed with mild moist climate, quite a contrast from the icy rigors of Labrador and northern Quebec at the same latitude on the other side of the Atlantic. The great Prussian and European naturalist and explorer Alexander von Humboldt noted that with the discovery of the New World, the observation of this difference between the eastern and western shores of the Atlantic constituted an important new contribution to the understanding of climate, until then focused on north-south contrasts.[3]

The Gulf Stream of the North Atlantic carries back east the waters that were driven to the west in the Tropics, and similar gyres dominate ocean circulation in the North Pacific with the Kuro-Shiyo, in the South Pacific, and in the South Atlantic oceans. Under the trade winds, the water heated to 30°C (86°F) reaches a western shore (an east coast) where it leaves the tropical zone going either north or south. It then moves to higher latitudes and crosses the ocean from west to east, finally bringing warmth and humidity to the west coast of another continent. There, part of the flow continues on to subpolar latitudes, while part turns back in the direction of the Tropics, becoming a cold west coast current. In these gigantic gyres flow millions of cubic meters of water per second (in fact, south of Nova Scotia, 150 million cubic meters per second, 750 times the Amazon River's flow). Near its start off Florida, Gulf Stream waters stand out clearly from neighboring ocean waters, in color as well as in temperature, flowing at several kilometers per hour (a few knots). At that speed, a bottle tossed in the water can cross the ocean in a few months. In contrast, the water hardly moves at all in certain areas within the gyre, in particular in the Sargasso Sea, where all sorts of debris accumulate.

Both the color and the temperature of seawater can be monitored from space, provided that no clouds block the view. The color of reflected sunlight gives information on biological activity—more precisely, on chlorophyll of living algae and on suspended sediments. It can be measured by radiometers with appropriate filters, a difficult measurement requiring that the weak reflection be distinguished from sunlight scattered by air molecules (the blue of the sky). Measurement of sea surface temperature (SST for specialists) depends on radiometers sensitive in "window" wavelengths of the infrared for which the atmosphere is transparent, and it also requires cloud-free skies. With such observations from space, scientists can keep track of phenomena such as the formation of cold-water and warm-water rings, eddies that spin off on each side of the Gulf Stream. Also, locating warm-water and cold-water fronts helps fishermen to go after certain species such as tuna that have particular temperature preferences.

Observation from space has revolutionized our knowledge of the world's oceans, but it cannot provide all the information needed to understand how the ocean works. Oceanographic measurement campaigns and *in situ* observations from instruments on buoys remain essential, and other types of observations can be imagined. No joke: a bottle can carry a message, and in the course of organized measurement campaigns of the last few decades, oceanographers have tossed well over 100,000 bottles into the sea, recovering 2 to 3 percent of them. And there have been a number of unplanned but welcome experiments as well. On May 27, 1990, the Korean containership *Hansa Carrier*, bound for the United States, met up with a huge wave and lost part of its cargo of Nike Air sports shoes. The containers washed away opened, releasing more than 60,000 shoes in the sea; but because of their air soles the shoes didn't sink, and after several months afloat they began to reach the shores of Canada and the northwest United States (and in fairly good shape, but as single shoes rather than as matched pairs). To take advantage of this unexpected bounty (some insurance company's loss), beachcombers organized exchanges so as to match left and right shoes. Once oceanographers heard of it, they set about collecting information on the dates and places where the shoes washed ashore. It was a bounty indeed: releasing 60,000 "tracers" at a single time and place would have been far too expensive an experiment, but here commerce and nature cooperated without asking for money from the oceanographers or their funding agencies. Another such unplanned experiment took place two years later, this time a cargo of plastic bathtub toys (floating ducks, frogs, and turtles) usually meant for baby's bath. These acts of God unwelcome to shippers and their insurers came as a maritime manna for oceanographers, providing enormous data on the details of the fine structure of the Kuro-Shiyo and other currents of the North Pacific. Who knows? Maybe some of the toys will succeed in crossing the Arctic and finally reach the North Atlantic. Some bottles released in the ocean at Nome, Alaska, in 1900 actually made it through the Bering Strait and reached the Irish and Norwegian coasts some ten years later. If you come across an interesting object on the beach, tell your favorite oceanographer about it. In addition to the major ocean gyres, all sorts of smaller-scale eddies, currents, and (mostly equatorial) countercurrents complicate the picture, not to mention the complexities of the third vertical dimension. We'll dive into that further on.

LAND AND SEA: MONSOONS

Sailors early learned to count on the reliability of the trade winds, but tropical seas do not stretch round the Earth. Interrupting the oceanic expanses, land masses complicate the simple picture drawn so far. In the course of the summer, land heats up more quickly than the oceans, this for a variety of reasons. First of all, the warming produced by absorption of solar radiation cannot penetrate far into the ground; it only affects a layer limited to several feet in depth. With a correspondingly small effective heat capacity, the land surface temperature varies strongly between night and day, even more between winter and summer. By comparison, tides and the winds agitate the liquid ocean, mixing the relatively thick surface layers. Only in extremely calm seas does the surface "skin" temperature rise significantly during the day to fall at night. Everywhere else, solar radiation must in effect warm a layer of water mixed down to at least 10 meters (33 ft.). In addition, especially where winds are strong, evaporation from the sea surface carries away latent heat, leaving less energy for warming the water. As a result, sea surface temperatures rarely rise above 30°C (86°F). At night, and during the winter, sea surface cooling proceeds slowly because, as the uppermost layer cools, the denser water sinks and is replaced by deeper not-so-cold and dense water. The resulting mixing of the surface layer can extend down as deep as 50 fathoms (300 ft. or nearly 100 m) and so involves enormous masses of water. As a result, variations of sea surface temperature are usually moderate and relatively slow.

On land, by contrast, nothing stirs up the soil: in the topmost few inches, temperature can climb strongly during the day, especially for dry land with no loss of latent heat by evaporation. As a result, summer daytime temperatures reach extremely high values, often more than 50°C (122°F) in the deserts of central Asia, the Sahara, Mexico, and the southwest United States. The high temperatures lead to the development of centers of lower atmospheric pressure, and the low-pressure centers attract cooler air masses from the neighboring oceans, where temperatures do not rise so much and atmospheric pressures remain higher. Over the Indian Ocean, the development of stronger and stronger contrast in surface temperature and pressure triggers the phenomenon called the *monsoon*, a name given by Arab navigators of the Middle Ages because the reversal in wind direction over the Arabian Sea actually reverses the flow in the Somali current east of Africa. The trade winds blowing from the southeast in the southern part of the Indian Ocean, drawn north of the equator by the low pressures, veer to the right because of the Coriolis force, and end up as southwesterly winds. The zone of towering clouds associated with rising warm humid air masses is driven by the monsoon winds further and further north of the equator, in the direction of the Indian subcontinent.

Reaching India, the thick water-laden clouds burst over the hills of the Western Ghats lining the coast near Bombay, but enough moisture remains in the air flow for monsoon rains to water the interior as far as Delhi and beyond. Once it reaches the foothills of the Himalayas, the moist air must rise and cool; and this produces a true deluge, with sometimes as much as 20 meters (780 in.) of rain falling at Cherrapunji in Assam. The great summer monsoon, the southwest monsoon, is a lifegiver, essential for food production for most of India. If the monsoon is good, everyone is happy and smiling, except perhaps for the European tourist who complains because it's raining in August. But late or weak monsoons are a catastrophe. Famine now can be avoided by assistance and commerce, transferring grain from areas with surpluses (mostly North America, Europe, and Australia) to those where crops have failed, but many problems remain. Hoarders and speculators may benefit more than those who need the food, import of "dumped" surpluses discourages local farmers, and grain is of use only if it can be transported without spoilage to the consumer.

Note that the Indians often speak of the "winter monsoon," referring to the winds that blow from the northeast in November and December, bringing rain to the southeast coast of India, in particular to the coast in the states of Madras and Pondicherry. These regions don't get much water from the summer monsoon, the southwesterly air flow losing most of its moisture as it crosses the highlands of the subcontinent. But is "monsoon" the right word? In fact, these are the northeasterly trade winds, part of the "normal" air flow of the Tropics, reinforced however by strong outflows of cold air from the Siberian winter anticyclone. Many specialists prefer to reserve the term *monsoon* to refer to the flow of warm moist air from the ocean, across the equator, toward the land, and this also occurs in West Africa, where the arrival of moisture in the low-level monsoon is crucial for survival in the Sahel. A somewhat similar situation occurs in North America, even though it is quite far from the equator. Warm moist air masses from the Gulf of Mexico bring rain to areas in the interior of the continent far from the Atlantic and separated from the Pacific by the high barriers of the coast ranges, the Sierra Nevada, and the Rockies. And on much smaller scales, contrasts between land and sea—specifically, stronger faster daytime warming and nighttime cooling of the land, and higher thermal inertia of the water masses—drive the daytime sea breeze and nighttime land breeze familiar to visitors to the seashore. Fair-weather cumulus clouds form as warming air rises over the land during the day, and dissipate at night when the warming ceases and air begins to subside. The opposite occurs over the sea. Similar alternation of cloudy and clear skies also occurs between lakes of large area and their surroundings. In Africa, Meteosat (infrared) images show Lake Victoria covered by clouds at night, clearing up in the day as clouds form over the surrounding land.

EAST-WEST: EL NIÑO, LA NIÑA, AND THE PACIFIC BASIN

The Pacific Ocean, with eleven time zones separating the shores of Chile and China, covers more than a third of the Earth's surface, spanning more than 7,000 miles at the equator. More than 7,000 miles west of the coast of Ecuador, the equatorial Pacific bathes the Indonesian archipelago (often called the "maritime continent"), and further west, the equatorial zone traverses the Indian Ocean. From the shores of East Africa to the coast of Ecuador and Peru, this enormous oceanic domain and, above it, the atmosphere partly blocked by the Andes Mountains, pulsate with El Niño and the Southern Oscillation. Why these two names? In fact, until the 1950s, oceanographers knew El Niño in eastern Pacific waters, and meteorologists knew the Southern Oscillation of atmospheric pressure over the southern equatorial and tropical Pacific, as separate interesting phenomena. Only starting around 1960 did they recognize that the two phenomena were in fact connected parts of a single wider-scale pulsation, recognizing more recently the distinctive "personality" of La Niña. Integration of oceanography and meteorology continues as observational tools are improved and coverage extended to the entire globe, as vast increases in computer power make possible the modeling of coupled atmospheric and oceanic processes.

Fishermen along the coasts of Peru and Ecuador have always known about El Niño. Every December, as they prepare to celebrate Christmas, the birthday of El Niño (the little boy, the infant Jesus), the trade winds tend to weaken, and warm waters take the place of missing upwelling cold water from the depths of the Pacific. Fish are harder to find. It's a good time to rest, repaint the boats, and enjoy the festivities of Christmas and the New Year. But sometimes the vacation drags on: the trade winds don't come back, coastal waters become warmer and warmer, fish rarer and rarer. For fishermen, such an El Niño warm event constitutes a calamity, and repercussions extend to prices in the world grain markets because of increasing dependence on fishmeal as an alternative to grain for livestock feed. El Niño also decimates the bird populations that feast on fish, and that reduces their droppings of *guano* on the rocky islets where they roost. In the nineteenth century, when these dejections constituted an important and valuable source of fertilizer, each El Niño event produced a steep increase in market price. With warm waters invading the coastal zone, rainfall over nearby usually arid land increases enormously. El Niño may be a disaster for the fishermen, but it brings a year of abundance (*año de abondancia*) to farmers, insofar as they can keep flooding under control. Written records of these dramatic events have been kept since the arrival of the Spanish conquistadors in the sixteenth century. Analysis of layers of snow and ice accumulation in the glaciers of the Andes shows that such strong fluctuations of precipitation have been going on for several thousand years at least.

It is fairly easy to understand how the warm waters of El Niño lead to reduction of fish stocks (at least of the commercially interesting species, anchovies in particular). In the early 1800s, the great explorer Alexander von Humboldt described the cold current flowing north along the Pacific coast of South America. Sometimes called the Humboldt current, but officially known as the Peru current ever since oceanographers decided no longer to use the names of persons, this is part of the great gyration of South Pacific ocean currents. Near the equator, surface waters are usually driven west by the easterly and southeasterly trade winds. Off the coast of South America, the westward drift has a pumping effect, producing upwelling of cold water from the ocean bottom to replace the warmer surface water. Thanks to this upwelling of bottom water rich in nutrients, marine life thrives: the nutrients fertilize the growth of phytoplankton or algae, the zooplankton (microscopic sea animals) graze in these marine meadows and are eaten in turn by fish, with birds and humans feeding on the fish. However, when the trade winds weaken, the upwelling dies down. But why should the trade winds weaken?

Between 1900 and 1930, many centuries after Peruvian fishermen first discovered El Niño, an English meteorologist, Sir Gilbert Walker, director of the Indian Meteorological Department, was trying to find a way to predict failures of the Indian monsoon. After an exhaustive study of measurements made at weather stations all around the Pacific basin, he established the existence of coherent variations of atmospheric properties over all of this vast domain, a phenomenon that he named the *Southern Oscillation*.[4] Near the equator and in the Tropics, whenever surface atmospheric pressure in the western Pacific (for example at Darwin, northern Australia) falls below its average value, higher-than-average pressures prevail over the eastern Pacific (Tahiti, or Easter Island); and vice versa. In the western equatorial Pacific, low pressures go with intense convective activity, strong ascending motion of warm moist air, towering cumulonimbus clouds, and high precipitation. Higher pressures correspond to much less of this, indeed to drought. A sort of seesaw of atmospheric pressure variations spans the vast domain of the equatorial Pacific, with effects extending still further east and west along the tropical belt, and also north and south over all the Pacific. Today we consider the phenomenon to be of global scale.

Let's not get ahead of ourselves! At the time, meteorologists were far from realizing all the implications of the Southern Oscillation. Even today, they are far from being able to produce reliable forecasts of the fluctuations of the Indian monsoon, and that was Walker's original objective. The east-west surface-pressure seesaw goes along with a shift of the area of strong convective activity from Indonesia in the west toward the central and eastern Pacific, and back again, interpreted in terms of what is now called the "Walker circulation." In the so-called "normal" situation, low atmospheric pressure, strong convective activity, and abundant rainfall are observed over Indonesia, while at the same time to the east atmospheric pressure is higher, cloud cover reduced, and hardly any rain falls at all. This goes together with strong trade winds blowing to the west near the surface, with return flow to the east at upper levels of the atmosphere, somewhat like a Hadley cell oriented east-west rather than north-south. This Walker circulation cell involves strong uplift of warm humid air and accompanying precipitation over the Indonesian archipelago, and subsidence of dry air over the central and eastern equatorial Pacific. The subsidence of dry air with virtually no rain is perfectly consistent with the extremely arid conditions that usually prevail over the few islands to be found in this part of the Pacific, as well as over the coast of Peru. In the other phase of the Southern Oscillation, with higher pressure to the west, flow in the Walker cell is reversed: subsidence and aridity over Indonesia, ascending air and anomalously high rainfall in the east.

It was only in 1957 that Jacob Bjerknes, a Norwegian meteorologist settled in California, recognized that El Niño on the one hand, and the Southern Oscillation on the other, were in fact two aspects of a single phenomenon involving strongly linked oceanic and atmospheric processes (fig. 7.3), today often called ENSO (El Niño/Southern Oscillation) by specialists. With pressures high to the east, the trade winds are stronger, vigorously driving the surface ocean waters to the west, thus activating the pumping and upwelling of cold bottom waters in the Peru current. Together with reduced evaporation from colder waters the subsidence of dry air practically rules out rainfall over coastal Peru and the eastern and central Pacific. Drifting to the west under the strong trade winds, warmer surface waters pile up in the western Pacific, and sea level there is several tens of centimeters (a few feet) higher than to the east.

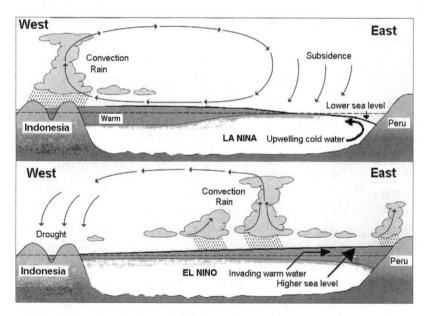

FIGURE 7.3 The Southern Oscillation: El Niño (*bottom*) and La Niña (*top*). The two panels show schematic east-west cross-sections of the Pacific Ocean and the atmosphere above it, along the equator, with vertical scales highly exaggerated. The thickness of the warm water layer (shaded) is only a small fraction of the depth of the ocean, and the tilt of the sea's surface, though real, is very small. Surface atmospheric pressure is above the average in the areas of subsidence and drought, below average in areas of convection and rain.

Sea-level measurements were once only possible along coasts, using tide gauges. Some such gauges have long been installed in Pacific island harbors, but there are precious few islands in the central and eastern Pacific. The advent of satellite radar altimeters has completely changed the situation, and beginning with the Franco-American Topex-Poseidon oceanography satellite launched in 1992, data on sea level are available over all of the oceans with an accuracy of a few centimeters. In the 1960s it had already been noted that sea-level changes between east and west tended to follow the Southern Oscillation's seesaw of atmospheric pressure, and the satellite data provide much more precise and comprehensive coverage. In the "normal" situation, with high pressure to the east, water warmer than 30°C (86°F) piles up in the west, blown there by the trade winds strengthened by the "attraction" of the lower pressures over Indonesia, the maritime continent. But at some point, this pileup can no longer continue. Triggered perhaps by a burst of westerly winds, a sort of giant wave of warm water moves off to the east, inexorably invading the eastern equatorial Pacific all the way to Peru. El Niño is on his way, and his extended stay in Peruvian coastal waters always corresponds to the phase of the Southern Oscillation with pressures higher in the west and lower in the east. There, the trade winds are weakened by the repulsive effect of the stronger pressures to the west, and weak westerly surface winds take their place. The layer of warm water invading the eastern equatorial Pacific prevents cold bottom water from reaching the surface while at the same time providing humidity to the local atmosphere by enhanced evaporation. The humidity added to the atmosphere condenses when the warm air rises, releasing latent heat, stimulating more rising motion and further lowering the surface pressure; and that reduction of surface pressure in the east further weakens the trade winds. The atmospheric and oceanic processes reinforce one another, and El Niño extends his stay. What is the cause, which is the effect? What triggers the El Niño event, and how and when will it come to an end? In fact, the same set of oceanic and atmospheric processes can operate in the other direction, restoring strong easterly trade winds, the warm water moving back west, and the resumption of the upwelling of cold bottom water along the coast.

Up to the end of the 1980s, the tendency was to consider the El Niño events as a sort of anomalous state, returning on average twice a decade, to be distinguished from the "normal" situation. However, the El Niño or ENSO warm event is part of an *oscillation*, the Southern Oscillation, with

El Niño as an extreme phase of that oscillation. What about the other extreme phase? In 1988, for example, a strong east-west difference of pressure was observed, with unusually high values to the east just off Peru, and temperatures at the surface of the eastern Pacific significantly lower than "normal." And it seems now well established that the shift of the atmospheric jet stream that accompanied this extreme ENSO phase (with cold water off Peru) played a major role in the extremely hot and dry conditions that prevailed in summer 1988 over much of the United States, which some had hastily attributed to "global warming." After some hesitation regarding the name to give to cold events of this type, the scientific community seems to have settled on the name of *La Niña* (the little girl).

As we shall see, the ENSO (or better still, ENLNSO?) phenomenon has strong impacts on regional ecology, the regional and world economy, and society. Are reliable forecasts of the extreme phases possible? There certainly is a need, and there has been significant progress in the research and operational weather service community. Following the very strong El Niño of 1982–83, major efforts were made to strengthen monitoring of the Pacific. A comprehensive network of buoys was set up and has since been operating, with some buoys drifting, some fixed, some making measurements at depth as well as at the sea's surface, all along the equatorial band of the Pacific, transmitting their data by way of communications satellites. At the same time, new tide gauges were put into operation at many additional sites, and new satellites have provided more and more data. An important first step in the research was to identify some advance signals of particularly strong El Niño or La Niña phases.[5] After all, the warm water mass takes several months to cross the Pacific, before the warm or cold event or "anomaly" reaches its peak. Does the "advance signal" tell us something about the *cause* of these phenomena? Is there in fact a cause to be identified? After all, in the ENSO phenomenon, one observes *oscillations* around an average state called "normal." On a year-to-year graph of the Southern Oscillation Index (the difference between atmospheric pressure in the east and in the west of the Pacific, e.g., between Darwin and Tahiti), the curve goes above and below the average value, sometimes more, sometimes less, sometimes hesitating, but it is virtually never steady. The combined ocean-atmosphere system of the Pacific basin is hardly ever in its *normal* state, if by that its average state is meant.

Let me propose another way to look at this oscillation. Let's go to the park. In the Parc Montsouris near my home in Paris, I occasionally see and hear Spanish-speaking children at play, because it's just across from the international student campus. In New York's Central Park or Riverside Park, near where I used to live, you can easily find Spanish-speaking children, a little boy and a little girl, El Niño and La Niña, together on a seesaw. You don't know how long they've been there. At one moment, El Niño is up, then La Niña. Sometimes they teeter a bit, nearly balanced in a horizontal position, but they never stay that way long; that may be the average state of the seesaw, but it's not normal in the sense that nearly half of the time one side is up or going up, the other down, and vice versa. Occasionally, La Niña will give herself a strong kick to get up high quickly; or maybe it's El Niño who gave the strong kick, or who shifted position closer or further away from the central axis. Sometimes, El Niño seems stuck longer on high. How can you explain all these ups and downs? You don't know who started things off, but between seesaw mechanics and child psychology, the kids keep going up and down. Of course, if one of them is much heavier than the other, a parent is needed. But just as for the seesaw in Central Park, for an overall explanation of the ENSO (ENLNSO) phenomenon, rather than look for an external cause, research tries to understand how the many interactions between atmospheric pressure, wind, sea level, atmospheric temperature and humidity, and water temperatures at different depths combine to keep the oscillation going. During the last ice age, 18,000 years ago, the ENSO seesaw didn't work quite the way it works today, but for the last 10,000 years El Niño and La Niña have been at it, up and down, a bit more, a bit less, a bit faster or more slowly.

This section started off with the El Niño warm events and the associated decimation of anchovy populations off the coast of Peru, but this is only a part of the enormous oceanic and atmospheric fluctuations affecting the entire Pacific basin and beyond. Reversal of the east-west atmospheric motions in the Walker cell necessarily affects circulation between the equator and the Tropics in the Hadley cell, with additional consequences beyond the Tropics, over the whole Pacific at least. In particular, the changes in the distribution of atmospheric pressure lead to the formation of more typhoons traveling to islands such as Tahiti and Hawaii normally not visited by such storms. The shift of the jet stream sends one winter storm after another in the direction of California, and with a little bad luck, luxury sea-view homes, built on slopes that are typically threatened at least once a decade, end up on the beach or in the sea below. Maybe the insurers will pay, but the rates will go up.

In Indonesia, extreme El Niño events bring drought, as in 1982–83 and 1997–98, and any fire, in particular those set to clear forests for palm oil plantations, can easily go out of control, burning vast areas and sending stifling smoke to pollute cities as much as a thousand miles away. In the eastern Pacific, suddenly abundant rains (ten times the average!) fall on usually desert islands, radically transforming the landscape as luxuriant vegetation quickly takes advantage of this water from heaven. It becomes a veritable deluge in Peru (3 m, i.e., 120 in. of rainfall, in six months!), Ecuador, and Colombia. Some farmers know how to profit from this (planting rice instead of cotton, for example), but the high humidity and flooding raise serious safety and health risks.[6] And further off? Walker was aiming at forecasts of Indian monsoon failures, and statistically, it seems that the monsoon is indeed weaker during the El Niño years. In any case, the effects of the changes in atmospheric circulation and ocean temperatures extend across the Indian Ocean to the East African coast, and depending on the season, drought afflicts the south and east of Africa from Ethiopia to Mozambique.

And still further away? The atmosphere knows no borders, so that everything is related, but that doesn't mean that all the "teleconnections" that can be imagined really exist or are significant if they do. In many cases, no chain of cause and effect can be established, although statistics suggest a link. But the world is round, and it would be astonishing if an upheaval on such a vast scale as ENSO were to stop at the Andes. In the tropical zone, when El Niño rains drench the western slope of the Andes, drought reigns to the east, in the Amazon basin of Brazil. The strong perturbation of the Hadley cell over the Pacific and Indian Oceans also has effects over the Atlantic, although it is by no means the only factor governing variations there. Does a strong El Niño constitute a global disaster? It's certainly a gigantic upheaval over and in the Pacific. Billions and billions of tons of warm water go on the move, the locations where water evaporates (and where now fresh water falls as rain) shift by thousands of miles, winds change direction, and all this happens in a matter of months. Elsewhere on the globe, less dramatic but nonetheless real and important changes are observed. Nevertheless, the catastrophe, in the mathematical meaning of a radical change of state, is not necessarily a worldwide calamity. No doubt El Niño hurts Peruvian fishermen, as it does those improvident Californians whose perilously perched houses slip into the sea. But for prepared Peruvian farmers, El Niño rains bring abundance. For the United States as a whole, the balance sheet must be worked out. The associated distribution of atmospheric pressure and the route taken by the jet stream tend to keep hurricanes away from the U.S. East Coast and the Caribbean, otherwise strongly under threat. Accountants tell us that the gains (or rather, the absence of losses) by residents of Florida and their insurance companies more than make up for losses in California, an El Niño benefit seldom mentioned by the media. Moreover, the El Niño weather teleconnections generally bring mild winters to the northeast United States. Cold-related deaths as well as heating fuel costs are reduced. During the exceptionally mild 1997–98 El Niño winter, in place of unneeded heating oil that could not be sold, refiners switched to producing and selling gasoline at bargain prices, encouraging still more gas-guzzling.

Despite the headlines, El Niño does not constitute a climate catastrophe everywhere in the world. What about La Niña, the opposite phase of the oscillation? Not much in the news before 1988, La Niña also has mixed effects, both boon and bane. It's fine for fishing off Peru and for fire-quenching rainfall in Indonesia, but drought afflicts farmers both in much of the United States and west of the Andes from Colombia to Peru. Moreover, while El Niño tends to keep hurricanes away from the U.S. East Coast, the Caribbean, and the Gulf of Mexico, La Niña brings more active hurricane seasons. Hurricane Mitch, the cause of many thousands of deaths in Honduras and Nicaragua, came in October 1998 *after* the end of the 1997–98 El Niño and a rapid switch to La Niña conditions.[7] The 1999 hurricane season was extremely active, but fortunately no hurricane hit such a vulnerable target. The media have helped to make the public aware of the worldwide impacts of El Niño (especially in 1997–98), but often they give the impression that the apocalypse has come. Far be it from me to play down the damage and pain brought by violent weather events—tornadoes, typhoons, hurricanes, or wild winter storms—but we have to recognize that the climate of the Earth works through ceaselessly changing weather, winds, and waves. Like it or not, we have to live with the weather, and ignorance of its violent side is no excuse. We know the statistics fairly well, and reliable forecasts of some though not all of the most dangerous events are increasingly available. For the media, however, the saying "No news is good news" seems often to have been transformed into "Good news—or no bad news—is no news."

NORTH-SOUTH: THE ATLANTIC

Especially in the atmosphere, the "waves" made by the Southern Oscillation spread out over the globe, but in Europe, on the opposite side of the world from the South Pacific, their effects can be detected only by sophisticated statistical analysis of changing weather patterns. In Europe, and also in northeastern North America, much more noticeable effects arise from what has long been known as the North Atlantic Oscillation (NAO). As with the Southern Oscillation, meteorologists observe a seesaw of atmospheric pressures and associated wind and water variations, but in this case between north and south rather than east and west. It shouldn't come as a surprise. Compared with the vast east-west spread of the Pacific, the Atlantic is squashed, especially between the Brazilian *nordeste* and West Africa, where only 1,800 miles separate the two shores. Furthermore, North Atlantic waters mix quite freely with water and ice from the Arctic, whereas the Bering Strait and the Aleutian Islands form a bottleneck between Arctic and North Pacific. In the subtropical North Atlantic, an area of high atmospheric pressure often known as the Azores anticyclone separates the zone of easterly trade winds in the Tropics, to the south, from the zone of prevailing westerly winds at northern midlatitudes. Sometimes the anticyclone spreads out and shifts in the direction of Europe, providing clear skies there while blocking weather systems arriving from the west. European vacation-planners look forward to the expansion of the Azores anticyclone, many farmers fear it. As for fishermen, they know perfectly well that they have to manage with the Iceland Low and its bad weather where Arctic air meets mild and moist midlatitude air.

Like everything else in the atmosphere, the Iceland Low and Azores High are never fixed, either in location or in intensity. Here too there is a sort of seesaw of atmospheric pressure, between atmospheric pressures over Greenland and Iceland in the north, and over the Azores Islands in the south, changing the strength of winds and the tracks of Atlantic storms. When the Iceland Low deepens and shifts north toward Greenland, when at the same time pressures over the Azores get still higher, the increased pressure difference strengthens the west winds over the midlatitude Atlantic, and also the trade winds blowing westward off the Sahara Desert over the tropical Atlantic (fig. 7.4). Additional desert dust reaches Caribbean islands such as Barbados and even Florida, and surface waters off the southeastern United States tend to be warmer than average. The stronger west winds further north pick up additional warmth and moisture from the Gulf Stream before reaching Europe. Exactly where rain falls and where skies are clear can vary, but generally this positive NAO phase involves mild and wet winters in western Europe.[8] However, if the Azores High spreads over France, conditions can be dry and cold in the winter, dry then becoming hot during the long sunny days in the summer. By contrast, in the negative phase of the NAO, the Iceland Low is pushed south and weakened by the development of high pressures over Greenland, and also pressures over the Azores are not quite so high. Then west winds over the Atlantic weaken, but with blocking by the Azores High weakened, they can bring more winter rain to southern Europe, the Mediterranean, and North Africa. Also in the negative NAO phase, extremely cold dry air from the Arctic can reach northern Europe, giving quite cool summers and bitter cold winters, sometimes as far south as France.

From one year to the next, the winds will vary, blowing warmer or colder as they exchange more or less heat and evaporate more or less moisture while passing over warmer or cooler Atlantic Ocean surface waters. Thanks to the Gulf Stream, the waters that wash the shores of Ireland and Brittany are far less chilling than those along the shores of Labrador and Newfoundland, at much the same latitude to the west. Thanks to the warmth that northeastern Atlantic waters give up to the atmosphere and that west winds carry to Europe, winter temperatures in Dublin, London, Paris, Brussels, and Geneva are far milder than those that are the common lot of citizens of Minneapolis, Montreal, or Moscow. But by the same token, western European city-dwellers cannot count on using the subway to get to cross-country skiing or open-air natural ice-skating, one of the pleasures associated with cold winters. North of Iceland, and sometimes also to the south, floating ice forms each winter and melts the following summer, but on occasion icebergs wander off to the south without immediately disappearing. Since the sinking of the *Titanic*, a systematic effort is made to keep track of such drifting hazards, and in recent years satellite radar has made it possible to monitor icebergs from space even when the sky is completely overcast. The limit of the floating ice pack advances and recedes, not only with the seasons but also from year to year. Indeed, careful study of the archives of maritime meteorology has revealed that in addition to the year-to-year fluctuations, there are strong variations from one

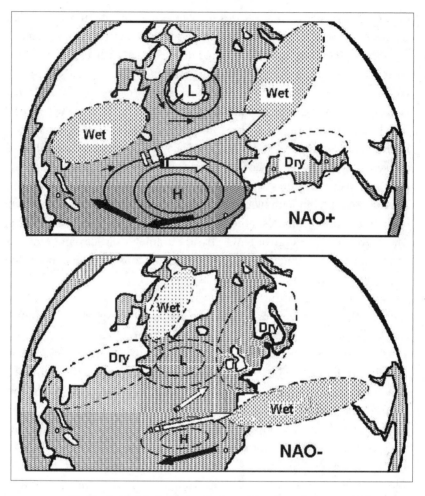

FIGURE 7.4 The North Atlantic Oscillation. In the positive phase, the stronger high-pressure center near the Azores and the northward shift of the low-pressure center from Iceland to Greenland strengthen both westerly winds at midlatitudes and the northeast trade winds in the Tropics. These accompany shifts in wet and dry areas, in warm and cold ocean currents, and in many other phenomena.

decade to another. In the Labrador Sea, maximum ice cover occurred in 1957, 1972, and 1984; and each time, the average North Atlantic sea surface temperature went through a maximum a year or two later. What can explain these variations, with warmer water following closely the maximum amount of ice? Oceanographers have some ideas but no certainties. Further complicating the issue, the North Atlantic Oscillation appears to be related to a north-south pressure seesaw between the North Pacific and northwestern North America (Alaska, northwestern Canada).[9] But anything happening over the Pacific is necessarily affected by the Southern Oscillation and, likewise, also has an effect on it. The Earth is round, and the atmosphere knows no borders.

These north-south variations, especially those that operate more slowly than the Southern Oscillation with its switch from El Niño to La Niña in a few years, complicate the problem of determining to what extent we are already seeing the global warming trend that must result from the reinforcement of the greenhouse effect. For one thing, weather satellites have existed only since 1960, and true global coverage was not possible before then. With only four decades of such global observations, and with strong NAO changes from one decade to the next (or even periods of twenty to thirty years), it is risky to say that the true long-term trend has been unambiguously extracted from the data. The importance of the Atlantic Ocean should not be underestimated, despite its being much smaller than the Pacific. For one thing, the Atlantic actually does export water to the continents, the Americas as well as Europe and Africa, whereas overall, the Pacific imports water, with slightly more rainfall than evaporation. Even more noteworthy, the Atlantic Ocean transports heat northward not just in the Northern Hemisphere starting in the Tropics, but all the way from the Southern Ocean, across the equator, up to Greenland and Scandinavia. To learn how this happens, we have to dive into the deep ocean.

CHAPTER 8

WATER'S DEEP MEMORIES

THE SALT OF THE SEA

Few scientists give credence to the idea that water has memory, if by that is understood memory at the scale of a single or a few H_2O molecules.[1] However, the great masses of water on Earth do undoubtedly have memory in the sense that they retain the mark of past environments. Ice contains most of the fresh water on our planet (table 6.1). As many as a million years of changing atmospheric composition and climate are recorded in the accumulations of snow changed to ice, and some 420,000 years of these "archives" have been deciphered.[2] Even a very small ice sample contains very large numbers of hydrogen and oxygen atoms, and the ratios of the different hydrogen and oxygen isotopes, measured at different depths in the ice, tell the story of past variations of atmospheric temperatures and global ice amounts. Winds of the past transported dust from ancient deserts to the polar caps and, falling with the snow (then covered by the following year's snow), the dust remains in the ice today. Even more remarkably, trapped air bubbles tell us what the atmosphere was like thousands—indeed hundreds of thousands—of years ago. Extracting ice cores from the 2-km (1.2 mi.) thick layers of ice covering Greenland and Antarctica, glaciologists retrieve a faithful record of past climates going back over 400,000 years. Mountain glaciers too consist of ice from accumulated snowflakes, some of which fell over 20,000 years ago. But the downhill flow of glaciers distorts the ice layers, complicating the analyses. Usually, the bodies of unlucky climbers and skiers fallen into crevasses turn up at the outflow not more than several decades later, but sometimes they only are discovered after a much longer delay, as in the case of the hunter Ötzi found virtually intact with his clothes, arrows, and other equipment, after several thousands of years buried by snow in the Tyrolean Alps. To the extent that the ice remains frozen for long periods, such water in solid form does have memory. Liquid fresh water does not, subject as it is to continual renewal, spending just a short time in the atmosphere or on the surface between condensation, precipitation, and evaporation, or falling and flowing into the salty sea. The only exception: water deep underground, such as "fossil" water under the Sahara or in the Ogallala aquifer of the western Great Plains since several thousand years, or much older and not so fresh water such as the brines found deep under the Paris area.

Some "memory" does exist in the oceans, which contain in liquid but salty form nearly all (97%) the water on our planet. Although renewal or rather mixing of this water takes place in times quite short compared with solid-Earth processes, it still takes hundreds of years or more, so that traces of the past are not immediately mixed up and wiped out. They do, however, require interpretation. Water of the top hundred fathoms of the ocean, refreshed in a few years, retains but little memory of its past. The surface layer loses water by evaporation, gains fresh water from rainfall and from the rivers that flow into the sea, and makes the circuit of the ocean basins in the wind-driven currents. Even so, ocean waters affected by a climate anomaly—for example, off Peru during an El Niño—retain some trace of those conditions when they arrive elsewhere (for example, off the coast of Japan) ten or twenty years later. To keep track of what is going on, oceanographers must plunge into the third dimension and study the deeper ocean layers. Surface layers exchange water in different forms with the atmosphere, with no immediate effects on the abyssal depths, but ultimately mixing does occur. Purely mechanical processes enter quickly into action. Along western coasts near the equator, off Peru and Ecuador in the Pacific except when El Niño comes, westward currents driven by the trade winds produce upwelling, in effect pumping deep water. Further away from the coast, east-west pressure differences in the water drive the equatorial countercurrent, flowing from west to east at a couple of hundred fathoms deep.

On a research vessel drifting westward in the surface current, a scientist fishing for fresh tuna will see her line stretched to the east. With upwelling of water at some ocean locations, there must be downwelling somewhere else. Such downwelling occurs whenever a water mass has greater density than the water beneath.

The density of seawater depends on its temperature and its salinity (salt content), with cold salty water being the densest, fresh (rain) water less dense, ice still less. On average, seawater contains about 35 grams of various salts per liter (of which 30 g of ordinary salt, sodium chloride NaCl), but the proportion varies. The atmospheric portion of the water cycle operates as a still, producing fresh water. Solar-powered evaporation removes H_2O molecules from the sea, leaving the salt behind dissolved in the water. Of course, some wind-borne droplets evaporate, leaving sodium chloride crystals carried off as salt spray by the wind, deposited here and there on boats, beaches, and inland. Wherever evaporation exceeds precipitation and river runoff, the sea surface layer gets saltier and saltier. In the Red Sea and the Mediterranean, salinity reaches 40 grams per liter. That may not seem to be much of a difference from 35, but as a result, Mediterranean water sinks at a meter per second once it finds its way over the (300-meter deep) threshold of the Strait of Gibraltar into the Atlantic. Although it slows down further on, this water of Mediterranean origin sinks to a depth of 1,000 meters, and it can be distinguished by its salinity all the way to the Gulf of Mexico. In a similar way, the super-salty waters of the Red Sea spill into the Arabian Sea. And where the water can't go anywhere but just evaporates under the Sun, as in the Dead Sea (Israel, Palestine, and Jordan) or Utah's Great Salt Lake, high salinity makes the brine so dense that it can support a swimmer (usually a tourist posing for a picture) without swimming.

In the tropical zones of the major ocean basins, abundant sunshine produces strong evaporation while heating the water. Surface waters of the western Pacific, spreading eastward during an El Niño event, reach temperatures of 30°C (86°F) or higher. But abundant rainfall keeps this "warm water pool" from becoming very salty, and the warm surface waters float above cooler denser waters. I've already mentioned that the Atlantic and Indian Oceans export water to the continents, losing more water by evaporation than they gain by precipitation. For the Pacific, the opposite holds, making its surface waters generally less salty than the Atlantic's. Away from the Tropics, as evaporation weakens, salinity drops. Lower salinity also holds near the coast, especially at the mouths of great rivers bringing fresh water back to the sea. With abundant influx from the Ganges and Irrawaddy Rivers, the Bay of Bengal, east of India, is far less salty than the Arabian Sea, receiving fresh water from the Indus River, but much less, and losing fresh water by strong evaporation all around Arabia, as in the warm and salty Red Sea and Persian Gulf.

Around the poles, precipitation and river influx bring more water to the oceans than they lose by evaporation, but another process—freezing—plays the key role in determining salinity. Fresh water freezes at 0°C (32°F), but depending on its salt content, seawater's freezing point can be 1 to 2°C (1.8–3.6°F) lower. Furthermore, when seawater freezes, the ice contains less salt than the liquid water left behind. If you melt sea ice floating in the Arctic Ocean, or on the edge of a very cold beach (not in France or even Scotland, but in northern Canada, Alaska, or Siberia), you won't get fresh water, but with only 10 to 15 grams of salt per liter, it will be a lot less salty than the seawater. When sea ice forms, it leaves very cold and very salty liquid water behind, water so dense that it sinks to the bottom of the sea. All around Antarctica, along the edges of the ice pack, under the ice shelves, and especially in the Weddell Sea, this process pours cold and very salty water into the bottom of the Southern Ocean. The resulting layer, called *Antarctic Bottom Water* (AABW), spreads out along the bottom, filling up the abyssal plains. Advancing north around the mid-ocean ridges, it keeps its salty "personality" as far as the equator and even beyond. The same process operates up north in a more concentrated pattern along the edges of the ice pack, between the Arctic and the North Atlantic (fig. 8.1). Some of these phenomena reverse when sea ice melts, as some of it does each summer, releasing a layer of less salty water. From 1968 to 1972 a large mass of such less-salty water drifted from the east coast of Greenland to Newfoundland and continued on with its distinct low salinity to the north of Scotland in 1976 and up to Spitsbergen in 1979. Interacting with the North Atlantic Oscillation, this "Great Salinity Anomaly" certainly had some influence on weather in western Europe and eastern Canada.

The dense cold salty water flows south along the sea floor, channeled by the continental shelves and the mid-ocean ridges into a powerful undersea stream. As with the spread of the Antarctic Bottom Water, the northern bottom waters keep their personality even south of the equator. Acoustic

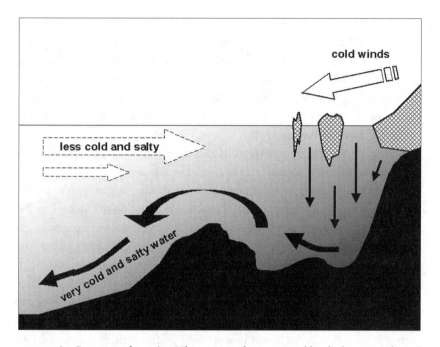

FIGURE 8.1 Deep-water formation. When seawater freezes, very cold and salty water is formed and sinks to the bottom of the sea, to be replaced by a drift of less dense and cold water from lower latitudes.

sounding of ocean depth began in the nineteenth century, and more or less detailed maps of the sea floor have existed for many years, although some results obtained by the United States and Soviet navies during the Cold War, especially in the cold oceans, were secret until very recently. Practically no deep-water passages link the Arctic and Pacific Oceans. On the Atlantic side, by contrast, the cold salty water pouring into the bottom of the Norwegian, Iceland, Greenland, and Barents Seas moves south, surging over an undersea ridge between Greenland and Scotland, flowing around the Faeroe Islands and Rockall[3] to become North Atlantic Deep Water (NADW). Not only does this outflow make the circulation of the Atlantic profoundly different from that of the Pacific, it actually affects the whole world ocean.

How do we know what currents are flowing in the depths of the seas? Again, one way is to toss a "bottle" into the water, but this time it must sink to follow the flow. Instead of bottles, oceanographers have learned to use various types of easily tracked tracer particles. Seawater contains dissolved atmospheric carbon dioxide, with some of the carbon in the form of carbon-14, produced naturally from nitrogen by cosmic rays reaching the upper atmosphere, and easy to identify because of its radioactivity. The proportion of carbon-14 has varied with time. Over the last century, this is in part because fossil fuel burning has released to the atmosphere a substantial amount of CO_2 containing practically no C-14. With a half-life less than 6,000 years, all of the C-14 that once existed in the organic material that ultimately became coal or oil decayed millions of years ago. Much more recently, between 1950 and 1964, some C-14 was produced by nuclear bomb testing in the atmosphere, and the propagation of that 15-year-short pulse can also be followed in the ocean's depths. The same goes for the tritium (hydrogen-3) produced by the explosions, partly incorporated in the H_2O of rainwater and then seawater, later decaying to form identifiable helium-3 that remains dissolved in the sea. Industrial production of chlorofluorocarbons (CFCs) began in 1938 and increased strongly in following decades, but was brought back down to nearly zero following the Montreal Protocol of 1987. These gases also dissolve in water, and no chemical reaction destroys them until they reach the stratosphere where they reinforce ozone depletion. Analysis of seawater at different depths shows that these different tracers, "bottles" tossed into the water for the most part only since the 1940s, have reached the bottom only near the poles, and not at all in the Tropics. Near Bermuda, tritium had reached depths no greater than 700 meters in 1970, but could be found at 2,000 meters in 1990. Off Brazil, at latitude 10° South, tritium is found in the topmost 700-meter layer; some traces have reached depths of 1,800 meters, none deeper. At high latitudes, on the other hand, the tritium "signal" coming from the atmosphere and the surface has reached bottom. This invasion of the lower depths does not necessarily proceed smoothly: southeast of Newfoundland, CFCs descended from 1,000 to 1,700 meters depth between 1983 and 1991, then sank to 2,000 meters, finally reaching bottom more than 4,500 meters (15,000 ft.) deep before 1995.

Following the tracers—for the most part chemical or radioactive pollutants—requires regular sampling of the water in different parts of the ocean. Tracking them as they drift toward lower latitudes in the deep sea, oceanographers can draw the three-dimensional map of the deep ocean current system, often called the *thermohaline* circulation because it depends on both temperature and salinity.[4] There are other ways to sound the seas. In the

Arctic, automatic measurement stations, put in place by air-supplied scientific expeditions, drift for years with the floating ice. Not only do they measure and relay standard weather data (temperature, pressure, wind in the near-surface air), they also monitor the state of the ice (several feet thick), and they measure physical, chemical, and even biological properties (temperature, electrical conductivity, flow speed, salinity, dissolved CO_2, plankton, etc.) of underlying seawater down to 100 meters (330 ft.) deep. Instruments to measure the water's drift are moored at depth in the strategic passages. In addition to instrument-carrying buoys drifting with surface currents, oceanographers have designed, built, and released hundreds of buoys drifting at constant depth. These automatic underwater observatories determine their positions by analysis of acoustic signals emitted from a number of known fixed points, and return to the surface from time to time to send the recorded data back to the oceanographers by satellite relay.

The heat content of an entire ocean basin can also be estimated, and not only with thermometers.[5] The speed of sound waves in water depends on the water's temperature as well as pressure and salinity, and at low frequencies (less than 600 hertz or cycles per second), these sound waves can travel enormous distances without any significant attenuation. Having set up a number of sound emitters and receivers around the western Mediterranean, a French-German-Greek research team used acoustic tomography to draw a three-dimensional map of this nearly closed basin between Gibraltar and Sicily, analogous to the way in which computer-aided tomography (CAT) X-ray scans are used to map the inside of the human body. Other teams have used acoustic sounding to study what goes on under the ice pack near the Melville Islands of the Canadian (Inuit) Far North. Many oceanographers would like to do the same for the entire North Pacific, but this project has so far been blocked by the concern that powerful low-frequency sound waves might be harmful (or at least bothersome) to great whales. Maybe the project can be carried out using lower power.

Much has been learned since the 1950s, thanks to unclassified deep-water research made possible by the development of bathyscaphes and other manned research submersibles in the United States (research submarine *Alvin* and others), Belgium (*FNRS*), France (*Cyana, Nautile*, and others), Italy (*Trieste*), Japan (*Dolphin 3K, Kaiko*), and elsewhere. During the Cold War, much was also learned but kept secret by the American and Soviet navies, as they explored beneath the Arctic ice and in the deep ocean, looking for the best places to hide nuclear missile submarines. Much more will be learned in coming years as the international scientific community gets to work with these data, now finally being declassified. Although there has been some tendency for military research funding to decrease, some formerly Soviet scientific research institutes are being kept alive by Western funding, and the new cooperation between Americans and Russians should yield more generally useful scientific results than the earlier separate conduct of secret military research.

THE GREAT CONVEYOR BELT

From the tropic of Cancer to the tropic of Capricorn, abundant sunshine—warming the surface and evaporating water—powers up the Hadley circulation, and this systematic overturning movement of the tropical atmosphere exerts its influence even at latitudes further north or south (fig. 7.1). Concentration and transport of solar energy in the form of latent heat (fig. 6.2) account for the violence of many meteorological phenomena. Below the surface, in the currents of the deep sea, the high-latitude "cold source" operates through the peculiar properties of freezing seawater to put into motion the planetary-scale overturning circulation channeled by the topography of the sea floor. Just as for the atmosphere, the Coriolis effect of Earth's rotation induces east-west motions in the ocean, this effect adding to the force exerted on the sea's surface by the winds. The sinking masses of cold salty water at high latitudes must be replaced, pulling in not-so-cold water from lower latitudes. As the sinking water leaves the Arctic zone to become North Atlantic Deep Water, the relatively narrow sea passages between Greenland, Iceland, and Scotland channel it into a powerful current of 14 million cubic meters of water per second, flowing south in the deep ocean between the continental shelf of the Americas and the mid-Atlantic ridge. Reaching the latitude of Cape Horn, there strengthened by Antarctic Bottom Water and caught up in the eastward circum-Antarctic flow, this mighty sea-floor stream brings cold and very salty water to the Indian and Pacific Oceans. There, it finally mixes with intermediate-level and surface water. While some of the water continues drifting east in the Southern Ocean, other water masses approach the Tropics, and at the surface they return west under the prevailing trade winds, warming under the Sun.

Caught up in the Agulhas current of the western Indian Ocean, and flowing around South Africa, these warmed waters make the long journey all the way to the North Atlantic, where they reinforce the Gulf Stream. This Atlantic warm water drift transports heat northward all the way from the Cape of Good Hope to the Norwegian Sea. Still lukewarm (or rather not very cold) with temperatures of 12 to 14°C (54–57°F) off the Newfoundland coast, this water gradually transfers heat and humidity to the atmosphere and, by way of the westerly winds, to much of Europe, maintaining a mild climate east of the North Atlantic. The water finally becomes truly cold in the Greenland and Norwegian Seas; some emerges in the Labrador Sea, but the coldest and saltiest water sinks back to the bottom, completing the giant loop and starting it over again. Overall, omitting many details— but remember, details are important, and not only locally—the waters at the surface of the Atlantic drift generally from south to north and bring heat with them, while at the bottom the current flows from north to south, taking "cold" along with it.[6] Oceanographers have named this combined planetary-scale system of surface and deep-sea currents linking the Arctic and Antarctic zones and all the world's oceans "the great ocean conveyor belt," which they term "conveyor-belt circulation" (fig. 8.2). The complete tour may take a thousand years or more. Somewhere at the bottom of the sea, there must be water that sank from the surface during the "Little Ice Age" three centuries ago, carrying with it traces (isotope ratios) of those colder climes. The ocean remembers.

FORESEEABLE ACCIDENTS, OR SURPRISES?

Anything that happens in the atmosphere—for example, radioactive fallout—makes itself fairly rapidly felt at the bottom of the ocean near the poles, but transmission of the "signal" through the whole ocean takes much longer. Does this mean that the ocean's "memory" and mass necessarily entail strong inertia of global climate, delaying any response to a sudden perturbation? Consulting the climate chronicles stored in polar ice, we find that we shouldn't count on it. Analysis of the ice core record over the last 100,000 years and more reveals several sudden climate jumps, strong temperature changes in less than a century, perhaps even in only a decade or so. Moreover, the fluctuations were nearly simultaneous in both polar

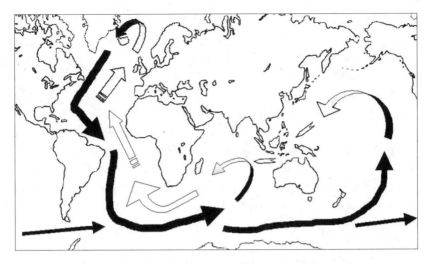

FIGURE 8.2 The great ocean conveyor belt, showing the deep ocean cold current (shaded belt) and the near surface drift (white arrows). The process illustrated in figure 8.1 forms North Atlantic Deep Water (NADW) between Greenland and Scandinavia in the north, and Antarctic Bottom Water (AABW) all around Antarctica in the south. Warming of the water finally returning to the surface occurs mainly in the Indian and Pacific Oceans.

zones and at intermediate latitudes. For the most part, these jumps took place before the current interglacial began, i.e., between a cold climate and a *very* cold climate. One exception stands out: the cold snap called the *younger Dryas*, about 11,000 years ago, at a time when the ice caps had already been in retreat for a few thousand years. Since the end of the younger Dryas about 10,000 years ago, climates have been relatively stable, fluctuating within a fairly narrow margin. Had they not been so stable, would humans have developed agriculture, or persevered in it?

Climate jumps are scary. (The French expression, *faire froid dans le dos*, appropriately evokes a shiver in your back, an icy fear.) What would happen in Europe, from France to Scandinavia, if temperatures were to plunge to values common in Nome (Alaska), with canals, rivers, and harbors icebound all winter, from October to May? Can we count on the stability of the conveyor-belt circulation? Uneasiness about this has developed since the discovery of the apparent coincidence of some of the cold crises with "Heinrich events." In 1988 the German oceanographer Hartmut Heinrich noted that debris found at the bottom of the sea off the Norwegian coast included rocks that must have been stripped by glaciers from the Canadian (Laurentian) Shield.

Armadas of icebergs calved from the ice cap must have carried the rocks across the Atlantic and finally dumped them next to Europe. If indeed such armadas of icebergs were involved, their melting must have released large masses of cold but relatively fresh water. This melt water, being less dense than the surrounding saltier water, would have formed a floating layer. By the same token, it would have blocked the sinking motion and so would have put a stop to the pulling in of warmer water from the south. Research groups in Germany and the United States have used relatively simple models to simulate this situation.[7] Results suggest that a substantial but not unreasonably large increase in freshwater input to certain ocean areas would lead to partial or total shutdown of the conveyor-belt circulation, perhaps even for a very long time, leading to much more freezing in Europe than has been seen for 9,000 years.

Plenty of controversy still surrounds the hypothesis of the iceberg armada as an explanation for the younger Dryas thousand-year cold snap. Still, the question raised by Wallace Broecker (1997) of Lamont-Doherty Earth Observatory of Columbia University needs to be taken seriously: by changing the composition of the atmosphere, are we humans running the *risk* of triggering a sudden revolution in the way the climate system works, either by blocking the conveyor-belt circulation or by radically rearranging it? Certainly the anthropogenic emissions of gases such as CO_2 tend to reinforce the greenhouse effect and so to warm the atmosphere. The consequences of such warming are still very hard to evaluate with regard to the changes in precipitation as well as to the accumulation of snow and ice and melting of ice caps, all of which determine freshwater flux to the ocean. There might be a bad surprise in store for us. In his book *The Imperative of Responsibility*, the philosopher Hans Jonas (1984) writes: "It is safer to pay heed to the prophet of doom than to the prophet of bliss." In recent years it's become quite fashionable to invoke the "precautionary principle," which has made its way into international law. However, can we really reason on the basis of unpredictable events? How can we distinguish between unacceptable risks and unreasonable nightmare fantasies? Should we stop everything until scientists can totally eliminate uncertainty (not likely!) and enlarge the domain of predictability? All paths to the future (and we can't help but go there!) are paths to the unknown, but it is plainly stupid to proceed without thinking, stepping on the gas in the fog, as if ignorance and uncertainty could eliminate risk.[8]

CHAPTER 9

CLOUDS, RAIN, AND ANGRY SKIES

MAKING RAIN

Water falls from the sky, but heaven's water is continually replenished by the cycle of evaporation and condensation. Covering more than 60 percent of the surface of our planet, clouds form by condensation. Water vapor, the invisible gaseous form of water, present in the air in variable amounts, becomes visible only when it changes to the liquid or solid state, in fog or clouds. Some clouds form at relatively low levels, at temperatures above the freezing point (0°C), but for others forming at very high altitudes, temperatures can drop below −40°C or even −70°C (−94°F). Look at the sky and enjoy the show! Clouds take all sorts of forms, some flat, others round, some arrayed in long lines or rippled by waves, some fragile floating veils, others bursting upwards in "a towered citadel, a pendant rock, a forkèd mountain, or blue promontory."[1] Over the years, professional meteorologists have elaborated a complete nomenclature of the different kinds of clouds, classifying them according to their shape, height, and state. Many of you who enjoy watching the skies know some of these terms.[2]

Stratus clouds, prototype of the general class of *stratiform* clouds stratified in relatively flat layers, thick or thin, are confined above and below by atmospheric layers where water vapor does not condense. Vertical motions are small, but the water vapor that condenses to form the cloud must rise to get there—except of course in the case of fog at the surface, essentially a zero-altitude stratus. Made up of liquid water droplets suspended in the air, stratus clouds can cover the sky with a nearly uniform shroud, appearing white from above by day and gray from below, reddish at night when illuminated by city lights. If the layer thickens sufficiently, ice crystals can form near the top and liquid water droplets coalesce into raindrops; these are the *nimbostratus* clouds.

By contrast, vertical motions play a strong role in the formation and appearance of the *cumulus* and related clouds, with their ever-changing shapes and towers. Made up of liquid water droplets, *cumuli* of varying size are generally found in relatively low layers (a few hundred to a couple of thousand meters): little fair-weather cumuli dotting the summer sky, with upward extensions often burgeoning on warm afternoons, cumulus *humilis* developing through the *mediocris* stage to become a cumulus *congestus* and bigger. Sometimes, especially in summer above the western Great Plains, they release thin curtains of rain that evaporate before they reach the ground. Some cumuli form at higher levels, in a structured deck we call *alto-cumulus*. Vertical convective motions with rising warm air and sinking cool air can also affect a layer of low-altitude clouds, giving it horizontal structure and thus transforming the stratus into *stratocumulus*, but not perturbing it enough for the clouds to break out of their confinement. When vertical convection strengthens, cumulus clouds grow dramatically, especially upwards. The cloud can reach and sometimes penetrate the tropopause at altitudes of 10 to 18 km,[3] becoming a spectacular full-blown *cumulonimbus* with a capping veil of ice crystals at the top, spreading out with the wind to form what is called an *anvil* (according to its shape). Such a cloud has then as its full name *cumulonimbus capillatus incus*. Cumulonimbus *thunderheads* nearly always give strong wind gusts and often drenching downpours. Sometimes, dark bulges—called *mamma* because of their breastlike shape—appear at the base of the cloud. And when a more threatening funnel shape looms, it can be prelude to a tornado. Harbingers of violent electrical storms, cumulonimbus clouds deserve respect.

Clouds of the *cirrus* family are constituted of ice crystals, and so they exist only at low temperatures, most often at quite high altitudes, 8 km at least. They appear in the form of filaments, sheets, commas (*cirrus uncinus*), and still other shapes. Sometimes they too are structured by vertical motions in a fairly thin speckled horizontal layer, giving *cirrocumulus*.

Sometimes the high-altitude ice crystal clouds cover the sky in flat sheets; and, if it is thin, such a *cirrostratus* veil produces a panoply of optical effects. These result from the basic hexagonal structure of all ice crystals, which can be arranged in plates, columns, or even more complicated structures. Each crystal acts as a prism, with a fixed angle of 22.5° between the directions of the entering and emergent light rays. When we see a halo—i.e., a circle of light with angular diameter 45° (half the angular separation between the zenith and the horizon)—we are seeing light rays that have come from the Sun (or the Moon, or Venus, since they also give easily observed halos) but have passed through tiny crystal ice plates floating in the air. If the air is quiet at the level of the ice, the tiny plates tend all to be oriented the same way, and more refracted sunlight comes from the parts of the halo at the same elevation but right and/or left of the Sun, giving *parhelia* or "sun dogs." Sometimes the floating ice crystals have the form of tiny thumbtacks, and they can produce the *circumzenithal arc*, a phenomenon that looks somewhat like a small inverted rainbow, sometimes quite bright, between the Sun and the zenith. Look for it on days with a high hazy cirrostratus layer and no low clouds between you and it. The phenomenon is not particularly rare, but you have to look upwards to see it.[4]

Rain and snow fall from the clouds, themselves the visible manifestation of water condensed into liquid droplets or solid ice crystals suspended in the air. But just how does the atmosphere make rain? In principle, water should condense when and where the temperature falls below the dew point, i.e., whenever the relative humidity of the air (relative to saturation) reaches 100 percent. In reality, condensation usually only takes place with some degree of *supersaturation* (relative humidity above 100%), and it depends on the presence of small particles suspended in the air—*aerosols*. Such particles act effectively as *cloud condensation nuclei* or CCN (not to be confused with CNN!), provided that they can be wetted (*not* the case of soot particles); ice crystals (falling from higher clouds) work very well, and when it's too warm for ice, various sorts of salt crystals, mineral dust particles, acid crystals, and other aerosols can do the job too.

Over the years, the hunt for an explanation of just how water vapor condenses and freezes in the atmosphere led many scientists on many merry chases—some after wild geese! Only in 1936 did Norwegian meteorologist Tor Bergeron come up with a hypothesis that seemed to fit the facts. In following years, Bernard Vonnegut, a physicist working at the General Electric

Research Laboratories in Schenectady, New York, fleshed out this hypothesis with numerous laboratory and field studies. He also must have influenced his brother Kurt, who in his novel *Cat's Cradle* imagined the ultimately catastrophic consequences of fabricating a very special (fictitious!) crystal ice structure. In the real atmosphere, ice crystals take many forms, all built around a hexagonal base: rods, plates, prisms, and the dendritic structures that produce the infinite kaleidoscopic diversity of snowflakes. But in low-level summer clouds, sometimes producing warm rain, no ice is to be found. Now to form raindrops simply by getting enough water vapor (H_2O) molecules together, without any other particles playing a role, would require an extremely high degree of supersaturation, never observed. In warm clouds, with no ice crystals available to act as condensation nuclei, other grains and crystals trigger condensation of water vapor, first in a thin film of liquid water around the particle, then with the tiny droplets (each nonetheless containing many many billions of H_2O molecules) coalescing to form bigger and bigger drops. These finally get big enough so that gravity's pull overcomes air resistance. For the smallest droplets (diameter less than 0.1 mm), the speed of fall is so slow as to be practically nil, so these droplets define the cloud. As the droplets grow in size, they fall more and more quickly: first drizzle appears, and with droplet diameters larger than 0.25 mm, rain falls. At the same time, the droplets can evaporate, smaller ones more quickly than larger, so that some clouds yield no rain at all, while the fine spray of rain falling from others, visible as trailing streaks, evaporates before reaching ground.

Cloud formation and rain depend to a large extent on the availability of condensation nuclei. Over land, there usually is no lack of particles, whether dust and other sorts of mineral or plant debris, or aerosols produced from organic molecules such as terpenes emitted by vegetation, as well as still others resulting from industrial or transportation-related pollution. All these make it relatively easy to form the water droplets of which low clouds are made. Over the sea, however, salt aerosols do not always provide enough CCNs. Over some areas, phytoplankton (algae) emit molecules of dimethyl sulfur (DMS) which then react to form sulfate aerosols, quite effective as CCNs for low clouds. Also, satellite observations often show "shiptracks"—brighter stripes of low cloud revealing the sulfur dioxide (SO_2) pollution due to use of high-sulfur fuel. Specialists now believe that industrial SO_2 pollution (especially when high-sulfur coal or oil is burned) increases the number of sulfate aerosols and significantly modifies the condensation process in low clouds. This doesn't necessarily mean that the amount of condensed liquid water changes: if extra CCNs lead to formation of more but smaller droplets, they enhance reflection of sunlight by the clouds.[5] That tends to cool the affected regions, or rather to reduce the amount of sunlight available for surface heating.[6]

Just as condensation of water droplets in low clouds depends on the availability of condensation nuclei, so freezing of ice crystals in high clouds requires that particles be present. To act effectively as freezing nuclei, their crystal structure must be compatible with that of ice. Because of these severe restrictions, freezing nuclei are relatively rare. Many clouds at temperatures ranging below the freezing point of 0°C down to −40°C (a typical value at an altitude of 8,000 m or 25,000 ft.) contain "supercooled" water—liquid droplets rather than ice crystals. But once appropriate aerosols are encountered (or introduced), freezing proceeds quickly, as the first tiny ice crystals formed act as so many freezing nuclei for the remaining supercooled water. Moreover, if ice crystals or snowflakes fall from a high cloud into an intermediate-level supercooled water cloud, there too the conversion to ice is extremely rapid. Such seeding plays a very important role in natural rainmaking. High in the atmosphere, at temperatures below −40°C, supercooled water practically never exists, and ice crystals form directly from water vapor without passing through the liquid phase. At these altitudes, the atmosphere is generally very dry, most water vapor having condensed at lower levels and then precipitated (i.e., returned to the surface as rain or snow). Nevertheless, some water arrives as vapor or ice at the cumulonimbus thunderheads, distributed from there in the upper levels of the troposphere by the winds. Kerosene-burning jet engines also add some water vapor and aerosols, triggering the formation of visible contrails. When upper-level humidity is close to saturation, the contrails stripe the sky, surviving for many minutes when winds are not too turbulent, otherwise spreading out to fill the sky with cirrus, especially around major centers of air traffic.[7] Under other conditions, the ice crystals evaporate almost immediately and the contrails are very short. Can jet cirrus change climate?

STABILITY OR INSTABILITY?

Humans have enjoyed and taken advantage of climate stability over the past 8,000 years at least, but considering that adding extra fresh water to the Norwegian Sea could block the deep-sea heat-conveyor belt, should we take such stability for granted? The general circulation of the atmosphere and oceans transports solar heat to the poles and water to the land. The operation of this heat and water redistribution system depends fundamentally on the unequal apportionment of insolation and the manifold instabilities of the atmosphere. Temperature generally decreases with altitude, basically because most conversion of solar radiation into heat takes place right at the Earth's surface, and because air cools when it expands as pressure (the weight of air above it) is reduced. However, if the difference between the surface temperature and the temperature at higher levels exceeds a certain critical value—in more technical terms, if the vertical temperature gradient becomes too steep—the atmosphere becomes unstable. Vertical motions set in, with warmer air rising and cooler air falling; this mixing of the air tends to reduce the contrast between the warmer temperatures near the surface and the cooler layers above. By the same token, if horizontal temperature contrasts become too great, other instabilities lead to mixing of the different air masses—cold dry air versus warm humid air—coming from different latitude zones. Such horizontal mixing requires horizontal motions, bringing into play the Coriolis force linked to the Earth's rotation. Furthermore, the cooling of warm humid air leads to condensation of water vapor and release of latent heat; in some cases, this can further destabilize the atmosphere.

Day after day, water is evaporated from most ocean surfaces and much of the land, and night after night too by winds blowing over warm seas. In contrast, precipitation is often concentrated in both time and space, varying strongly day by day and hour by hour even over small distances. Usual values of evaporation are several millimeters per day (say an inch or two a week); but several centimeters of rain can fall in a day or even an hour, and sometimes much more. Acting over several days, evaporation transfers water from large surface areas to the atmosphere, and only when the moist air masses rise and converge can rain (and snow) fall concentrated over a relatively small area in a short time. Thus a system of rain-producing clouds must draw in the moist air masses from afar, and the Coriolis force necessarily acts on these motions. Concentration of atmospheric flows of moisture goes along with concentration of rotational motions, and this accounts for the violence of so many weather phenomena.

From Fair-weather Cumulus to Cumulonimbus: Convection and Precipitation

Ground temperature varies from place to place, depending on incoming sunshine and on the properties of the soil (or pavement) and plant cover (if any). The very-large-scale variation of insolation and temperature between the tropics of Cancer and Capricorn gives rise to the Hadley circulation (fig. 7.1); on the local scale, such variations provoke convective updrafts, forming simple cumulus clouds. Warm moist air rises in updrafts to the level at which temperatures are low enough for water to condense and form the cloud, with downdrafts of dryer and cooler air between clouds. Sometimes, depending on the wind, convection organizes itself in parallel lines of clouds stretching over long distances. In 1935, taking advantage of the updrafts between the clouds and the wind along them, a German glider pilot flew all the way from Bayreuth in Bavaria to Brno in Czechoslovakia, a distance of 500 km. When instability reigns only in the lowest atmospheric levels, the clouds confined there usually disappear after sunset, when the surface begins to cool and evaporation ceases. The opposite tends to occur over the ocean, especially in subtropical high-pressure zones. Clouds develop at night in the cooler air above the still warm (or not so cold) ocean waters, with individual cumulus clouds arranged in linear arrays or in hexagonal honeycomb networks, sometimes developing into a continuous blanket of stratus. Warming up after sunrise, cloud droplets evaporate and the clouds tend to dissipate as the day progresses.

When the ground warms strongly and the temperature difference between the surface and higher air layers becomes particularly strong, atmospheric instability appears, reinforcing the rising and converging motions of warm humid air. The condensation forming the cumulus reaches higher and higher in the atmosphere, and the corresponding release of latent heat can revive the rising motions. With cumulus towers and pinnacles penetrating colder and colder atmospheric layers, freezing comes into play and, once ice crystals form, they act as nuclei, further accelerating freezing and condensation. From *cumulus humilis* to *cumulus congestus*, the cloud may finally grow into a true *cumulonimbus* (fig. 9.1), producing *hydrometeors*, generally rain and hail. Vigorous updrafts engender strong wind gusts. Hailstones form where supercooled water freezes around ice crystals, and they grow bigger and bigger when the updrafts drag them up and let them fall again through the supercooled water. In as little as ten minutes, the storm produces drenching downpours. At the base of the cumulonimbus, some of the falling rain evaporates, cooling the air as it drags it downwards in particularly strong downdrafts. The air in the downdraft must go somewhere on reaching the ground, and it spreads out in turbulent air-rings, with small-scale violent wind gusts or *microbursts* that are particularly dangerous for aviation. At the top of the cumulonimbus, often at tropopause level or even higher, ice crystals dragged off by the wind form the *anvil*. The violent updrafts and downdrafts (often as strong as 30 m per second, i.e., 67 mph) can separate electric charges of atmospheric molecules, producing high voltages that finally are discharged in lightning between the cloud and the ground or between different parts of the cloud. Scientists do not yet fully understand the vast variety of atmospheric electrical phenomena, sometimes taking mysterious forms (St. Elmo's fire, ball lightning, sprites, etc.).[8] At any rate, a lightning flash usually consists of several extremely rapid short-duration discharges with currents of tens of thousands of amperes; as they pass through the air, they heat it all at once to about 30,000°C, creating a veritable explosion, the thunder that you hear a few seconds after seeing the flash (if you hear it immediately, it's too close for comfort or safety!). Buildings will burn if the lightning current heats flammable material; thanks to Benjamin Franklin's invention of the lightning rod in 1752, many historical monuments that would otherwise have burned are still standing.

CYCLONES AND CONVECTIVE CLOUD SYSTEMS

An isolated local thunderstorm goes through its life cycle in less than an hour, leaving behind only upper-level cirrus blown off from its anvil and perhaps some altostratus at intermediate levels. A moving cumulonimbus can renew itself with a supply of moisture from the ground. In addition to isolated individual cumulonimbus clouds, there exist much more powerful convective systems, including squall lines, convective cloud clusters, and tropical cyclones. The isolated cumulonimbus grows because of convective instability alone, but other larger-scale mechanisms come into play for the bigger complexes. The Coriolis force, interacting with convection, concentrates the rotational motions of the atmosphere

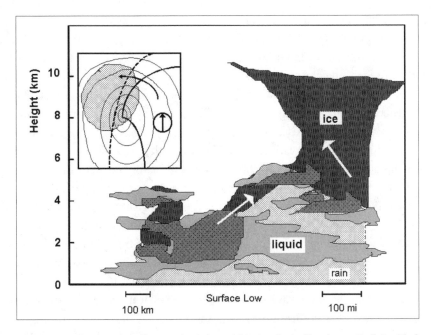

FIGURE 9.1 Cross-section of a cumulonimbus within a cyclonic disturbance. Both ice (*dark gray*) and liquid water (*light gray*) are present, with some parts mixed (*medium gray*); precipitation includes rain below and ice crystals (snow) above. Warm humid air is forced upward (white arrows). The cross-section follows the dotted line in the inset weather map (with arrow pointing north). The isobars (lines of equal atmospheric pressure) surround the surface low-pressure center, with a cold front to the northeast, a warm front to the south, and an area of rainfall to the northwest.

and so intensifies the violence of the winds. Whereas an individual storm affects only an area a few miles across, a line of active storms can stretch over dozens of miles, supplying itself with moisture from much larger areas. The system moves across the landscape with one or more convective cells vigorously drawing in and up the humid air and converting the water vapor into rain and hail at levels as high as 15 kilometers, all the while expelling dryer air from the cloud's top as well as in cold downdrafts spreading out at the ground. As the system advances, warm moist air ahead of it is forced above the descending cold outflow and so pulled into the convective circuit. And as the system is drawing in warm moist air from far and wide, the Coriolis force, acting everywhere except near the equator, gives the whole system a cyclonic spin, counterclockwise in the Northern Hemisphere, clockwise in the Southern (figs. 7.2 and 9.1).

Such systems develop at middle latitudes (35°–65°) as well as in the sub-tropical zone.

Atmospheric instabilities appear on the one hand when the fall of temperature with height becomes too steep (typically more than 6°C per kilometer or 3.3°F per thousand feet), on the other hand when horizontal differences become large. In the middle latitudes, especially in winter, strong north-south differences in insolation necessarily engender strong variation of temperature with latitude, leading to both horizontal and vertical changes in wind speed. With the Coriolis force acting on all of this, a simple flow pattern is unstable and cannot be sustained. Small perturbations of temperature and pressure necessarily grow and give rise to a cyclonic circulation around low-pressure centers, radically different from a well-behaved arrangement of temperatures and winds. Strong horizontal temperature and moisture contrasts—warm moist air masses versus cold dry air—appear across fronts (cold fronts, warm fronts, occluded fronts). Such small perturbations always appear, especially when the mostly westerly air flow must either go around or over mountain barriers. Thus the Rocky Mountains of North America and the Himalayas of Asia trigger an endless series of disturbances that become cyclonic systems as they cross the continents and the Pacific and Atlantic Oceans. In the United States, cold dry air masses coming in from Canada meet up with warm humid air from the Gulf of Mexico, and extremely strong instabilities develop along the front where they meet. In Europe, such meetings tend to be less dramatic, but to the north they occur very often, and indeed the theory of fronts and air masses was developed by Norwegian meteorologists during the first half of the twentieth century. In the winter, the ground is too cold to sustain convection, but strong north-south contrasts lead to the development of storms with both wind and snow.

In the midlatitude North Atlantic, the Gulf Stream acts as a source of warmth and humidity, while frigid air pours out of Greenland and off the icy waters of the Labrador and Greenland Seas. Such strong contrasts powerfully reinvigorate even weak low-pressure centers, perturbations, and fronts arriving from the North American continent. The perturbations cross the Atlantic in two or three days, and on occasion they strengthen explosively, becoming what some meteorologists actually call "bombs." Pressure drops (from 1,000 down to 990, 980, 970 millibars, or even lower),[9] and warm moist air is drawn from east and south into this low-pressure center. It must go somewhere and so it rises; and with the fall in temperature, moisture condenses releasing latent heat, and that heat sustains or reinforces the rising motion. Cold dry air descends on the other side of the low-pressure center. Picking up, northeaster wind speeds can reach 40 knots (40 naut. mph, 20.6 m, or 68 ft. per second—more than 70 km per hour or 42 statute mph), even 60 or 80 knots, up to force 12 on the Beaufort scale, with gusts of 90 or 100 knots. Wind speeds as low as 2 meters per second make waves; at 5 m/s (about 10 knots), whitecaps appear on the sea's surface. As the storm strengthens, and as long as it lasts, wave heights grow to 10, 20, and even 30 meters, extremely dangerous seas for trawlers and other fishing boats, with freighters at risk of capsizing, and even big ocean liners in trouble. When they reach the shore, the waves and the sea swell become a storm tide—not just a tide due to the gravitational pull of Moon and Sun with further sea-level rise due to the fall in atmospheric pressure but, most important, a rise in sea level because of the accumulated force of the wind acting on the waves. The storm tide floods low-lying land, sweeps away beaches, beach houses, and all but the strongest structures—an extremely dangerous event. Inland, such windstorms can still cause great damage, even when they aren't hurricanes.

On October 15, 1987, such a storm hit the coasts of England and France, ripping a big ferry from its mooring and pushing it aground further down the coast. On land, it knocked down several of the magnificent centuries-old oaks of the town of Seven Oaks (but it hasn't changed its name), demolished the glass houses of Brighton, and across the Channel in Trouville (Normandy), it blew off some of the roof tiles of the hotel where I and other members of the French and (then still) Soviet delegations were staying for a meeting on their cooperative program in space research. Winds and downed trees caused many accidents—some serious—among those foolhardy enough to drive in such wild weather.[10] More recently, early on the Sunday morning after Christmas 1999, the northern half of France was hit by an exceptionally violent storm with hurricane-force winds not only along the coast but also inland, across Paris and as far east as Germany and Switzerland. The weather service did forecast strong winds, but not that strong, and it was fortunate that most people were still in bed and not out on the street or on the road. A second violent storm hit southern France the day after, but most people were scared enough to stay inside, although lives were still lost. Between the two storms, power, telephones, and trains were out of service for many people, in some cases for many days. (The insurance companies are still paying and have raised their rates.) The wounds inflicted on the forests (millions of trees broken or uprooted!) will not disappear for many years, even decades. Although the media frenzy following the storm did eventually die down—albeit revived from time to time on the occasion of other weather events—the interest of the French public for the weather has not flagged since then.

Farther away from a North Atlantic storm center, a growing sea swell signals the action of a distant storm, sometimes the combined effect of several storms. Both on the American and European sides of the Atlantic, the shape of the shoreline and state of the beaches and rocks reflect the cumulative effect of the swell. The same phenomena occur on other seas. In the Southern Hemisphere, the Andes, the New Zealand Alps, and the hills of Australia and of South Africa disturb the midlatitude westerly air flow. Terrible cyclonic storms travel from west to east over the uninterrupted expanses of the Southern Ocean, in the roaring forties, the furious fifties, and the screaming sixties, where frigid Antarctic air meets warm moist air from zones closer to the equator. Those storms stimulate a powerful sea swell felt as far as the coast of West Africa; and in the Pacific, from Australia to California by way of Hawaii, the big waves break on dozens of surfers' paradises.

TORNADOES

Other storms grow over land. What with convective instability due to the steep fall of temperature with height above heated ground, and the instability of westerly air flow between markedly different conditions north and south, an extremely dangerous convective "super-cell" can develop, sometimes concentrating nearly all the Coriolis-induced atmospheric rotation in the area into a single particularly powerful vortex, sometimes into two. The low-pressure center draws in near-surface warm air from the southeast, while cold air flows out at higher levels to the northeast and southwest. While the super-cell moves eastward at speeds of 50 to 120 km/h (30 to 70 mph), whirling motion in the eddies intensifies and abates, often going through several cycles of growth, decay, and regeneration. Along a particularly active front with extremely high temperature and humidity contrasts, several super-cells may be in action at the same time. The activity can take the form of violent hailstorms, severely damaging crops and roofs. Worse, one or several *tornadoes* may form near to the center of the cell, often in a matter of minutes. The characteristic funnel-shaped cloud appears black mainly from the mass of debris sucked into it, and wind speeds higher than 60 meters per second (135 mph) have been measured at 200 meters from the center of the vortex. They are believed to be at least twice as large closer to the center, but instruments don't survive there. Thus it has never been possible to measure directly the pressure in the center of the tornado; but considering the weight of heavy objects (cows, cars, and trucks) observed to have been sucked in and tossed around, it must be at least 100 millibars below normal values. The tornado leaves a path of total destruction, fortunately seldom very broad or very long; but an active front can generate dozens of tornadoes. Although tornadoes do their damage mostly in America and Australia, they can strike elsewhere, even in western Europe. I was witness to a small tornado that removed part of my father-in-law's roof near Annemasse (Haute-Savoie, France), before uprooting several century-old trees in nearby Geneva, across the border in Switzerland. In the United States, tornadoes strike most of the time between the Rockies and the Appalachians, especially over the sparsely populated central Great Plains from Texas to Nebraska, along what is called "tornado alley." An extremely violent tornado outbreak associated with multiple supercell thunderstorms struck Oklahoma and Kansas on May 3, 1999, destroying or damaging thousands of homes, with the death toll reaching forty-nine. At Moore, a suburb of Oklahoma City, a truck-mounted Doppler radar of the University of Oklahoma measured winds of 318 mph, the highest ever recorded.

However, tornadoes can strike outside tornado alley, and at times they sweep through the southeast, where with greater population density and many flimsy or mobile homes, damage, injury, and loss of life can be much worse. This southward tornado track appears to occur more frequently during El Niño years, and early in 1998 several powerful tornadoes wreaked havoc in cities and towns between Tennessee and Virginia, including Birmingham, Alabama, and Nashville, killing over one hundred people and leaving enormous damage. During an earlier strong El Niño year, on March 18, 1925, a tornado left a 200-mile-long trail of terror and destruction from Missouri to Illinois, with 689 deaths and 11,000 homeless. Still, other deadly tornadoes strike when El Niño is not around. On May 31, 1985, an outbreak of some forty tornadoes left seventy-five dead from Ohio to Pennsylvania.

How do we know exactly what happens inside a tornado? Its narrow swath only rarely sweeps over a weather station, and when it does it hardly leaves the instruments in working condition (even if you can find them). Useful information comes from analysis of how the tornado has moved heavy equipment (trucks, construction equipment) of known weight, and from the damage inflicted on buildings whose construction has been well documented. Although they seldom yield details of individual tornadoes, geostationary weather satellite observations[11] make it possible to keep track of the weather conditions that spawn tornadoes. While up to now, routine operation provides full-Earth-disk images every half an hour, the satellite's instrument operations are modified when necessary to provide rapid (10-minute) scans of areas where dangerous rapidly evolving situations have developed. However, the most probing data come from the weather radar network because the emitted radar signal can penetrate the cloud some distance before the raindrops (liquid water) reflect it, signal penetration and reflection depending on the frequency used. The strength and delay of the return signal give information on the amount of water inside the system, and the frequency (or wavelength) shift gives a measure of wind speeds inside. On some occasions, very detailed measurements have been possible, using a mobile weather radar station brought close to the tornado's track. On the Great Plains, both professionals (in particular from the Severe Storm Center of Oklahoma State University in Stillwater and NOAA's National Severe Storm Laboratory at Norman, Oklahoma) and many amateur storm chasers eagerly await tornado conditions, bringing back spectacular close-up photos, videos, and recordings (of the storm's roar and their emotions), provided they don't get too close for safety (forget about comfort!).

With numerical models applying the well-known laws of mechanics, some scientists try to simulate on a computer what happens inside a tornado, while some experimentalists try to create miniature tornadoes, not in a test tube but in fairly small laboratory test chambers. Meteorologists have some understanding of how tornadoes are born and grow, but they have little hope of making precise and detailed advance forecasts. At best, generally dangerous conditions can be predicted a day or two in advance; then, keeping track of how the weather changes, areas of greater risk can be localized and more precise tornado warnings given with a few hours' notice; and finally, pinpointed tornado alerts are issued once the tornadoes actually appear. In the United States, such warnings and alerts appear automatically on television screens and are announced on the radio in directly threatened areas, on all stations and channels, but there still seem to be some people who have other things to do (work, read, talk to children) and leave the TV off. The U.S. National Weather Service has therefore developed and sells a special radio receiver providing weather data only, turning itself on automatically if there is a severe weather or flash flood alert for the specific local area. Provided that false alarms are kept to a minimum, such equipment will not end up in the trashcan and it may save lives. Still, when a tornado strikes, you need a solid shelter. The notice that I saw in Madison, Wisconsin, was to lie down in the bathtub if I didn't have time to get to the motel's shelter.

TROPICAL CYCLONES: HURRICANES AND TYPHOONS

Powerful convective systems also travel the Tropics. Between June and September, the low-level West African monsoon carries moist air from the warm sea surface in the southwest, across the intertropical convergence zone separating the two Hadley cells, to the African continent to the northeast. At higher levels, atmospheric waves—triggered perhaps when the African easterly jetstream-like air flow crosses the highlands of East Africa—travel from east to west, growing into clusters of convective clouds that supply nearly all the rain that falls in the Sahel. The trade wind air flow (fig. 7.1) then carries them off to the west, across the tropical Atlantic where sea surface temperatures exceed 27°C (80°F) before the end of northern summer. Abundant sun powers strong evaporation from the warm waters. Condensation of water vapor to form cumulus clouds releases solar energy stored as latent heat, amplifying the convective updrafts sucking more and more moisture-laden air from the lower atmospheric layers, this in turn maintaining evaporation from the sea's surface. The traveling rotating disturbance grows bigger and bigger, spanning hundreds of miles, while atmospheric pressure falls by tens of millibars in its center.

Such a tropical cyclone is called a *hurricane* in the Atlantic-Caribbean sector (after the Caribbean evil spirit *Hurakan*), once its winds exceed 60 knots (110 km/h, 72 mph). Winds in the enormous spinning system can strengthen further: in the biggest hurricanes, they can reach 300 km/h (180 mph) at about a dozen miles from the center. Walls of cloud grow higher and higher, releasing torrents of rain. The moist air entering the gigantic vortex does not reach its center; instead, dried-out air emerges at the tops of the towering convective cloud wall, and then descends to form the "eye" of the hurricane. Hurricanes in the Atlantic, such tropical cyclones are called *typhoons* in the western Pacific, but whatever their name, these monster storms can cause enormous damage with their winds, waves, and torrential rains, although at the same time the rain they bring is often essential for the water balance of the regions affected. Approaching the American coasts and the Caribbean, some hurricanes continue across the Gulf of Mexico to Texas, Mexico, or Central America, but others veer to the right, to the north, where the warm waters of the Gulf Stream continue to supply them with energy and moisture. They can reach the U.S. coast from Texas to Florida and anywhere along the East Coast up to Massachusetts. In August 1954 hurricane *Carol* swept across Boston, blowing down the Old North Church's bell tower from which the warning of the arrival of British troops was given, "On the eighteenth of April, in seventy-five" (the night from April 18 to 19, 1775), as Henry Wadsworth Longfellow told it in his *Ballad of Paul Revere*. In the same summer of 1954, two weeks later, *Edna* flooded roads from New York to Boston, and in October *Hazel* left over 1,000 dead. Much worse damage (700 dead, 63,000 homeless) was done by the unnamed hurricane—called by some the "Long Island Express"—that swept through Long Island and New England on September 21, 1938, with gusts of 186 mph measured near Boston, and peak wave heights of 50 feet at Gloucester, Massachusetts. Still, in most cases, the prevailing westerly winds of the middle latitudes push the storms out to sea before they can reach the coast. Note that since 1979 gender equality in hurricane naming has supplanted the era of *femmes fatales* that began in 1953. In September 1989, Guadeloupe had the misfortune of being visited by *Hugo*; in 1992, *Andrew* came very close to the center of Miami; and at the end of October 1998, *Mitch* left a terrible trail of death and destruction in Honduras and Nicaragua. The names of tropical cyclones are chosen in advance for each major ocean sector; the fourth Atlantic hurricane of the 2002 hurricane season will be named *Dolly*, with *Danny* the fourth in 2003, and *Danielle* back in 2004 after a five-year absence.

The greatest threat of hurricanes comes from the sea because winds of 100, 200, 300 km/h (as much as 200 mph) drive storm "tides" with the sea rising as much as 10 meters (33 ft.) or more, causing truly catastrophic floods. In September 1900, hurricane-driven floods overran the city of Galveston, Texas, leaving many thousands dead, and most survivors homeless.[12] Around the Gulf of Mexico, and from Florida to New England, vulnerable islands and beaches can be and are evacuated when necessary (and even sometimes when not), but in the Ganges River delta and along the shores of the Bay of Bengal, tens of millions of people live right at the water level. In 1970 a cyclone drowned hundreds of thousands of people in Bangladesh; and despite improvements in forecasting, another cyclone killed tens of thousands in India's nearby Orissa state in 1999.

Once above the continent, without warm water for regeneration, tropical cyclones lose much of their strength. But when the land itself is saturated with water, possibly brought by a previous cyclone, strong winds can cause damage far inland. Nowadays, we know hurricanes much better, thanks to the data from satellites and radar; thanks also to the daredevil pilots of the Hurricane Command, who take their planes right into the wild winds to measure the storm's vital signs. Still, numerical weather-prediction models can only include a very crude representation of what goes on in a hurricane, and forecasters have not so far been very successful in predicting, more than a couple of days in advance, the twists, turns, and spasms of these monsters. Nevertheless, tropical cyclones move slowly enough for effective warning to be given to directly threatened populations, time enough (at least in rich countries) to evacuate those most at risk and even to some extent to protect property (boarding up windows, etc.), provided that the warnings are taken seriously. In 1969 many of Camille's 259 victims were people who refused to leave their threatened homes on the Louisiana and Mississippi gulf coast.

Gently Falling Snow and Rain

No doubt, heavenly fury—wild and windy weather—brings water in torrents, but who can forget long days of soft rain, or those silent white winter mornings, all sounds muffled, as snow continues to fall? But even such gentle precipitation almost always falls from clouds that are part of a cyclonic system. After all, the supply of cloud water depends on the uplift of moist air, even when the clouds from which the rain or snow falls are apparently quiet stratiform clouds without violent vertical motions. While the storm rages near the center of a cyclonic disturbance, over hundreds of miles around this center (fig. 9.1), strong updrafts and the gentle rising motion of moist air must compensate the downdrafts and subsidence of colder dryer air, with the moist air reaching condensation and freezing levels of the atmosphere. Weather maps show cold and warm fronts as curves, but in fact they exist in three dimensions as more or less gently inclined surfaces rather than as vertical walls. The cyclonic disturbance leaves behind it a zone of nimbostratus trailing over at least 100 km, from which snow can fall for hours in winter, or rain when it's not so cold. Sometimes several cloud layers are piled up. Ice crystals may form in the topmost layer and, falling through clouds of supercooled water, they trigger further freezing in assemblages that become snowflakes. Below, in warmer cloud layers, the snow melts, giving rain; but in winter, if it's cold enough, no melting occurs and snow reaches the ground. If it's too cold for the air to hold much moisture, the snowflakes remain small and rare. Meteorologists describe this as a "seeder-feeder" process, transforming into precipitation 50, 60, or even 80 percent of the atmospheric humidity entering the system.

Both seeding and feeding are needed to make rain. A heavily water-laden cloud may well burst a bit sooner if "salted" with some appropriate particles (silver iodide crystals in particular). Can rain be made to fall in one place, and avoided in another? Can hailstorms be forestalled by advance rainmaking? For half a century, "rainmakers" have tried (often successfully) to sell their services, in particular to groups of farmers (or, during the Vietnam War, to the U.S. Department of Defense). In September 1987, Russian colleagues told me that the Soviet Hydrometeorological Service had organized flights of cloud-seeding planes in a large circle around Moscow to keep rain away from the city's festival, but if this was so, it certainly didn't keep us dry. Some meteorologists still believe that deliberate weather modification works, but serious statistical evaluations have never shown a significant effect. One can always seed the clouds, but if they can't be fed with water, nothing much beyond a short shower can be expected. When air currents do bring abundant moisture to sufficiently cold layers, Mother Nature usually seems to supply enough in the way of condensation nuclei for the impact of the artificial seeding to be undetectable.[13]

ACID RAIN

Pollution! It's not what it used to be: in London, Paris, or Pittsburgh, the air we breathe is no longer black with coal dust and soot, and atmospheric acidity attacks facades much less than it did during the 1960s. Pollution is not *where* it used to be either, some of it having moved to Athens, Mexico City, and Shanghai. In fact, in many places, pollution has only changed its color. The black pollution by soot due to incomplete combustion of coal has disappeared, letting through the Sun's rays and producing *brown* pollution by *photochemistry*, i.e., the complicated chemical reactions involving sunlight's action on incompletely burned hydrocarbons, in particular from the tailpipes of cars stuck in traffic jams. We're not necessarily breathing any better, and the ozone produced by such pollution damages vegetation along with our lungs. Furthermore, when industry made major efforts to reduce soot and smoke pollution, the priority was often to get results immediately visible to nearby neighbors, with emissions simply being spread in less visible form over larger areas by way of tall smokestacks. Pollution, a curse of industrial cities, moved out to the country. Toward the end of the 1960s, Scandinavians were dismayed to discover that their lakes were becoming more and more acidic, and complained bitterly of the "export" of pollution (mainly sulfur dioxide, SO_2) by the highly industrialized countries of western and central Europe, from Great Britain to Poland by way of the two Germanies of those years. It was only after the (West) Germans began to worry about the damage to and dying out of their forests (*Waldsterben*) that the European Economic Community and the Organization for Economic Cooperation and Development began to accept some responsibilities in the question of transborder pollution. Similar problems arose between Canada and the United States, leading to the signature of a treaty with the aim of reducing SO_2 emissions. However, after numerous studies, it appears that the question of forest "die-out" is not so simple— for example, nitrogen oxide emissions, accused along with SO_2 pollution of being bad for trees, actually may have a fertilizing effect (although they do human health no good). In any case, it seems reasonable to try to bring back down the concentration of sulfuric acid (H_2SO_4) in rainwater.

Did this only become a problem in the twentieth century? Certainly not. Right from the beginning of the Industrial Revolution, people from the countryside came to live crammed in mining and mill towns, in Darmstadt, Glasgow, and Manchester, with unbreathable air and "dark satanic mills" spewing black smoke and hiding the Sun. Over a hundred years ago, in 1872, Robert Angus Smith, royal inspector general for alkaline manufacture in Great Britain, published his book *Air and Rain*, describing his research on air pollution, his measurements of soot, of the acidity of rain, and of the increase in the atmospheric concentration of carbon dioxide (CO_2). For CO_2, his measurements showed that, compared with the countryside, CO_2 was higher by a factor 10 in poorly ventilated London theaters, by a factor 80 in the most insalubrious English coal mines. He even found that between the start and the end of a chemistry lecture at the Sorbonne in Paris, CO_2 concentration rose by 50 percent in relative terms, starting from an already high value. And as for soot, Smith made the following remark, still relevant today: "It is seen that the manufacturer's interest is not to make black smoke, because it is expensive, the public are interested in its removal, because it adds an unwholesome gas to those gases and solids already contained in ordinary smoke . . . and entails an enormous waste of heat."[14] And returning to the question of acid rain, Smith noted that generally rain is not acid far from cities, but that its acidity increases enormously in their vicinity. Between Valentia Island (off the west coast of Ireland) and Liverpool, he measured an increase by a factor 4 in the sulfates contained in rain, that factor reaching 16 in Manchester and 25 in Glasgow. Smith also noted how stone facades tended to crumble in the acid bath. Today, things have changed, at least in democratic advanced societies where citizens can do something to stop the acid showers.[15] Even so, pollution reduction often still means only making it less visible in urban areas and diluting it by spreading it wider.

CHAPTER 10

EARTH'S WATER, BETWEEN SKY AND SEA

WATERSHEDS AND WATER FLOWS

Water falls from the skies, and all open-air creatures large and small depend on its fleeting or prolonged passage over the land. Source and stuff of all life on our planet, water evaporates from the sea to return to the sea. But the water cycle does not follow a single track, and its many ramifications involve byways, bottlenecks, detours, and dead ends. Of the rain falling on land, two-thirds return rapidly to the atmosphere, evaporating in a matter of hours directly from soil and leaf surfaces, more slowly over days or weeks from rivers and lakes (table 6.1 and fig. 6.1). By contrast, water falling as snow may accumulate for months on the plains, for years, centuries, or millennia on high mountains or in the polar zones. Thanks to these slower stages in the race of water to return to the sea, crops growing in dry summer weather can still pump the precipitation of last winter, or even of previous wetter years. And thanks to the prolonged imprisonment of water in polar and mountain ice, twentieth- and 21st-century scientists have been able to study climates of yesteryear, indeed of hundreds of thousands of years ago.

At any moment, the world's streams and rivers contain only 2,000 cubic kilometers of freshwater (each cubic kilometer being a billion cubic meters or metric tons). But considering the total stream flow of water in all those streams to the sea (about 40,000 cu. km per year; see fig. 6.1), they must empty themselves and be renewed about twenty times every year. On average, water's "residence time" in rivers and streams is thus about eighteen days. The average can be misleading: fortunately, most rivers do not dry up after a month of drought. Lakes contain some fifty times more of the world's freshwater than do the rivers themselves, and when, as for the Nile, large lakes store part of the river's water supply, they provide a reserve for the dry season. In North America, some of the water entering the Saint Lawrence River spends as much as 300 years in the Great Lakes. These inland freshwater seas contain about 60,000 cu. km of water, a substantial share of the total freshwater in liquid form on the Earth's surface. Similarly, water spends 400 years in the depths of Siberia's Lake Baikal (23,000 cu. km), before it joins (at 53 cu. km per year) the Angara River flowing to the Arctic Ocean. In western Europe, the 38 cu. km of water in Lake Leman (Lake of Geneva) between France and Switzerland correspond to 250 days of Rhone River flow (partly maintained by the melting of Alpine glaciers); further north, Lake Ladoga with 908 cu. km of freshwater stocks twelve years' worth of the Neva River, only 74 km long, flowing through the beautiful city of Saint Petersburg (Russia) into the Gulf of Finland.

Lakes play a less important role for most other major rivers; or at least they used to, in an earlier natural state. Dam-building has created reservoirs and artificial lakes along many rivers lacking natural lakes, the aim being in part to regulate the river's flow, limiting flooding when rainfall is overabundant, providing water for irrigation in dry spells. Such human interference in the natural water cycle began several thousand years ago, but even today, over most of the continents, maintenance of some degree of regularity in stream and river flow depends on underground water gushing or seeping from springs, after months or years in the subsurface water table or deeper aquifers. Some underground water has been there a very long time, thousands of years and more. However, isotopic analyses confirm earlier suspicions that in most of the water table tapped by wells, the water is "meteoric," i.e., that it is fairly recent water from heaven even if it isn't yesterday's rain. The total amount of underground water (several millions of cubic kilometers) acts as a reservoir guaranteeing the outflow of streams and rivers (70,000 cu. km per year), and the underground detour of infiltrated precipitation can take decades, centuries, or even millennia. Wells drilled to irrigate farm or pasture land constitute an artificial shortcut taking underground water back up to the surface and atmosphere. However, even if expediting such water to the surface puts it back in the global above-surface water cycle, there is no reason to expect it to return rapidly to the area of extraction. In most cases, such water is lost for the aquifer from which it comes, either being evaporated or transpired by plants directly to the atmosphere, or running off into streams and rivers flowing to the sea. Pumping such water generally amounts to "mining" a resource that will not be renewed for a century or two at least. When the aquifer is not adequately recharged, such overexploitation may ultimately lead to desertification, as observed in many areas from the western Great Plains (where the Ogallala aquifer is being pumped dry) to the *sahels*[1] north and south of the Sahara Desert.

Average river flow rates and water residence times tell only part of the story; rainfall and snowfall vary from day to day and from place to place. The flow of water depends on widely differing conditions: gradual snowmelt in the plains of the Canadian or Siberian far north, violent local downpours producing flash floods in arid or semiarid areas, cloudbursts and thunderstorms in the tropical jungle or in the vineyards of southern France, showers in St. Louis, drizzle in Seattle, sleet in Boston. Depending on the climate zone, average annual rainfall ranges from practically zero in the Atacama Desert of northern Chile to as much as 3,000 mm (118 in.) in many tropical regions as well as along the temperate-zone Pacific shores of Canada and southern Chile. Moreover, rainfall varies strongly with the seasons in the Tropics, not so much in the temperate zones. Paris usually gets between 400 and 800 mm of rain in a year, between 40 and 70 mm every month. In 1985, a fairly dry year, less than 30 mm fell in February and October, but more than 60 mm in March, May, and June—519 mm altogether. Rainfall was over three times average in July 2000, and very high also from September 2000 to April 2001. New York City usually gets between 800 and 1,400 mm of precipitation in the year, with March, July, and especially August (start of hurricane season!), wetter than other months. But although there is neither a dry season nor a true rainy season in New York, there are dry and wet years. Things are quite different in the Tropics. In Niamey (Niger Republic, in the Sahel zone of Africa), *all* of 1985 was extremely dry, but 615 mm of rain fell in 1989. Nevertheless, less than 30 mm of rain fell in each month of 1989 except during the rainy season, when July (125 mm), August (303), and September (148) accounted for 93 percent of the year's total. Moreover, even though annual rainfall at Niamey is about the same as in the forest of Fontainebleau outside Paris, it can't support the same plant cover, for one thing because it all falls in a few months, for another because with higher year-round temperatures (especially during the dry season), water evaporates much more quickly.

In many regions, a single or maybe two or three events bring practically all the rain that falls in a month. An isolated rainstorm lasting hardly more than an hour may water an area only 10 to 40 miles square. A cyclonic system may remain active for a few days, bringing rain to much bigger areas as it moves, but at any particular place, rain—especially strong rain—may last only a few hours. In Europe, a solitary storm lasting only an hour may soak Amiens (north of Paris) one day, another may drench Brussels, the same day or the next. In southern France, it may not rain often, but when a storm sits for hours over Nîmes or Vaison-la-Romaine, serious flooding results. And when a large cyclonic system spends two days traveling across western Europe from England through Germany and Poland, the duration and intensity of precipitation are not the same in Plymouth (England), Cherbourg, Cologne, or Dresden, and not everyone may feel the same need to use an umbrella. In France, it rains only 10 percent of the time considering every hour; but in an average year in Paris, 75 days have more than 2.5 mm (0.1 in.) of rainfall, while there are 112 days with rain in Brest (Brittany), and only 49 in Nice on the Riviera. In the United States, New York City has 78 rainy days in an average year, Denver 37, Astoria (Oregon) 139, and Yuma (Arizona) only 12. In the east, cyclonic disturbances move up along the Appalachians from Georgia to Vermont, satisfying farmers with welcome water but sometimes feeding floods that wash away streamside homes, depending on the location and amount of the rain.

In some regions, topography systematically produces rainfall—for example, the Himalayas, the Bismarck Mountains of Papua New Guinea, the Pacific Coast Range of Canada and the northwest United States, the cliffs of the northeast coast of the big island of Hawaii. Elsewhere, however, although rainfall is often highly concentrated, those downpours don't always drench the same places. Once rain has reached the ground, or once snow melts, the water follows three processes: *evaporation* (or *evapotranspiration*) returns water directly to the atmosphere as water vapor, the gas; *infiltration* takes water down into the soil and subsurface layers, where it helps to regulate flow of water at the surface and at the same time affects the chemical properties of soil; and *runoff*, water moving mostly horizontally

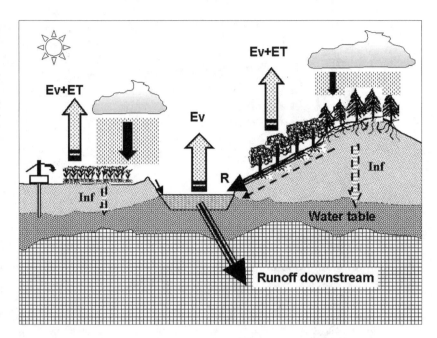

FIGURE 10.1 The water cycle on land. Division of precipitation (water from heaven) between evaporation (Ev) and evapotranspiration (ET), infiltration (Inf), and runoff (R). See also figure 6.1.

on downward sloping surfaces, collecting liquid water and taking it back to the ocean (fig. 10.1). Rainwater on a ridge or other line dividing the waters may make two (or more) very different journeys, depending on exactly where the raindrops fall. On a map, as well as on the terrain, those lines divide the land surface into *watersheds*. Provided it does not return directly to the atmosphere, water falling on the watershed will either infiltrate the soil or flow off on the surface; and most of the surface streams of a watershed join together to form bigger and bigger streams partly also supplied with water having seeped underground and out again. From gullies to rills to brooks to rivers, the water runs downhill and the streams run together. Half of the world's river flow to the sea is carried by the seventy biggest river systems, and seven of these account for more than a fourth of the total: the Amazon, Paraná-La Plata, Congo, Chang Jiang (Yangtze Kiang), Brahmaputra, Ganges, and Mississippi-Missouri (fig. 10.2). In a sense, diverting river water for irrigation constitutes a local delaying action against

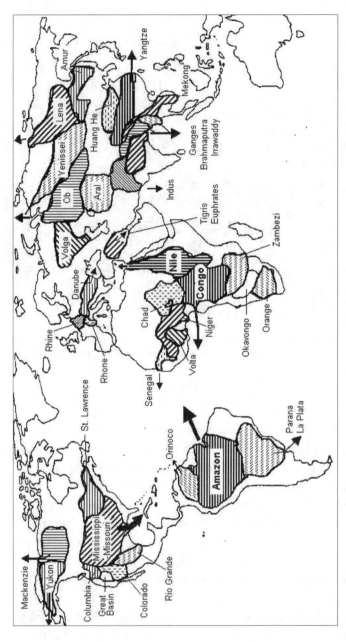

FIGURE 10.2 The great river basins. Note that Lake Chad (Africa), the Okavongo (Botswana, southern Africa), the Great Basin with Utah's Great Salt Lake, the Caspian Sea (into which flows the Volga), and what's left of the Aral Sea (into which flow the Syr Darya and the Amur Darya Rivers from Central Asia), all drain large areas but have no outlet to the oceans, losing water essentially by evaporation. By contrast, the very large runoff to the Pacific from Indonesia and New Guinea goes by way of a multitude of streams, none of which drains a large area.

the concentration of river flow, as it waters the land surfaces near (and not so near) to the river. Irrigation also strongly increases return of water to the atmosphere by way of the evapotranspiration of extended denser plant cover, and by evaporative losses from reservoirs, irrigation canals, and sprinkling.

VEGETATION, EVAPOTRANSPIRATION, AND THE SOIL

Plants make up nearly all the living matter on "dry" land, and water makes up the major part of that matter. Although their total water content comprises but a tiny fraction (much less than 1%) of the water available on land surfaces, plants play a key role in the division of water between evaporation on the one hand, and runoff and infiltration on the other. First of all, by way of sunlight-powered photosynthesis, vegetation uses part of the available water, together with carbon dioxide in the air, to make its own organic matter. The plant also needs water, in its sap, to transport other nutrients from the soil to the growing leaves. Oxygen (O_2) appears as a by-product of vegetation's manufacture of organic molecules from water (H_2O) and carbon dioxide (CO_2), and the exchange of CO_2 and oxygen (O_2) between the plant and the atmosphere necessarily involves some loss of water to the atmosphere: *transpiration* (in practice, evapotranspiration, as the transpired water evaporates). Roots constitute water-supply structures with which plants pump from the soil the water that is essential to their growth. The different gases (in particular CO_2) exchanged between the plant and the atmosphere pass through other structures, the small openings or pores called *stomata* distributed on the leaf's surfaces. The same stomata eliminate excess water pumped by the roots, depending on the temperatures of the leaf and of the air (and on the air's humidity). As it evaporates, transpired water carries off latent heat, keeping the plant cool. When the soil at the surface is dry, the plant suffers and begins to wilt, and its survival under such stress depends on its ability to pump less easily accessible water deeper down, assuming that any exists.

Plants thus return to the atmosphere some of the water that infiltrates the soil (fig. 10.1); depending on their needs and their root structures, they extract that water from layers as deep as a few feet or sometimes a few meters. In a single day, a linden tree can transpire 500 liters of water, a stand of oak about 30 cubic meters per hectare (12 tons per acre). The subsurface water can only evaporate once it is brought to the surface, and vegetation thus increases the evaporation rate relative to that of bare soil. In fact, evaporation from a surface depends on the availability of water there, its temperature, and on the temperature and humidity of the air in contact with the surface. Evaporation also depends on the wind speed because the air's movement carries off some of the water vapor in the immediate vicinity and so keeps the air next to the leaf's surface from getting too moist. On bare soil, the dried-out crust tends to block moisture deeper in the soil and so restricts evaporation. In a forest, leaf and needle surfaces abound, each one carrying stomata from which water pumped up from the roots can be transpired. Photosynthesis uses the energy in sunlight, but what if the Sun heats the plant too much? If it gets too warm, if the air is too dry, the plant can lose too much water by transpiration and dry out. The defense mechanism of certain types of plants (cacti in particular) is to keep their stomata closed during the day, opening them only at night. Other plants close their stomata whenever temperature gets too high. This is why you should avoid watering in the midday Sun. Evaporation of sprinkled water from the leaf's surfaces cools them, and the stomata open up: the leaves then get "burned," drying out before the added water gets to them by way of the soil, roots, and sap in the stem. And much of the water falling on the hot soil is wasted, evaporating before it has a chance to reach the roots.

In addition to pumping moisture from the soil and sending it back to the atmosphere by transpiration, vegetation, with its sometimes multitudinous leaf surfaces between earth and sky, enormously multiplies the chances of intercepting falling rain. Some drops remain on the leaves, and part of the water can return directly to the atmosphere by ordinary evaporation. Even if the water on the leaves soon drips to the ground, its temporary retention protects the soil from erosion during violent downpours, and tends to increase infiltration at the expense of runoff. The tree canopy also can provide some shelter to hikers and cows, albeit a dangerous one during an electrical storm. Of course, infiltration also depends on properties of the soil and the subsurface layers. But the soil, soaking up water and providing nutrients for the vegetation, is itself a product of that vegetation, as slowly or quickly decomposing plant debris is progressively transformed into *humus*. The topmost soil layer affects the chemical reactions between rainwater (or melting snow) and minerals and rocks deeper down, determining how weathering proceeds. At the same time, plant roots pierce what amount to veritable tunnels in the soil, and the various creatures that thrive around vegetation—worms, insects, rodents, weasels, and other animals—develop a labyrinth of passages that enhance still further the soil's permeability.

Weathering, or chemical erosion of rocks by the rain—one more example of pollution? Not necessarily! All of nature is chemistry. All rain can be acid. Raindrops are not pure water, because they result from the coalescence of cloud droplets, themselves forming around occasionally acidic cloud condensation nuclei. And raindrops fall through an atmosphere that has always contained some carbon dioxide. The CO_2 dissolves in the water, forming carbolic acid (H_2CO_3), furnishing ions of hydrogen and bicarbonate. Despite being relatively weak, carbolic acid attacks many different minerals. Granite, made up of quartz, feldspar, mica, and minerals containing silicates, aluminum, potassium, etc., is usually thought of as a hard crystalline rock. However, H_2CO_3 progressively turns it into various sorts of clays, among them kaolinite (or kaolin, a white clay from Kao-ling or Gao-ling, meaning "white hill" or "high ridge" in Chinese, a location above the Yellow River in Jiangxi province, where the Chinese extracted raw material for their porcelain as early as the seventh century A.D.), montmorillonite (after Montmorillon, a town in central France), and bauxite (the aluminum ore named after Les Baux de Provence in southern France). On granitic rocks, the "pure" rainwater filtered through the chemically active layer of humus, teeming with life (including bacteria and mushrooms), acts to form a layer of clayey soil containing fragments of quartz and other insoluble minerals. In many soils, clay plays a protective role, retaining water when it is abundant, giving it up when conditions are dry.

Elsewhere, under other skies and climes, with different plant cover, with other rocks, depending on whether the humus is completely or partly soaked or on the contrary dry and airy, weathering works in many different ways. In often-flooded bog land, with little ventilation of the leaf litter, decomposition and transformation into humus proceed only very slowly if at all. In marl, sometimes called "white clay," the soil is a sort of compromise between clays and limestone. Elsewhere, sand (silicon dioxide) dominates. In cold regions (for example, in the Siberian taiga), decomposition of organic matter proceeds very slowly, but the resinous forest litter makes the infiltrating water highly acidic, forming what are called *podzols* (after the Russian word for ashen soils). In still other areas, depending on the amount of iron oxides and hydroxides, of clays rich in silica, of aluminum oxides, and of organic matter in the soil, the earth can be red, as in the Deep South of the United States, or in many areas around the Mediterranean, or else brown or even black, as in the case of *chernozem* (literally, "black earth" in the Russian and Ukrainian *chernozium*).[2] In the humid Tropics, by contrast, frequent strong rainfall leaches out all the soluble minerals, including most of the silicates, leaving a brick-red soil containing fragments of quartz and insoluble oxides of iron and aluminum, with only a thin layer of humus surviving above it. So long as thick plant cover protects such soils, they maintain and are maintained by luxuriant living matter. But all too often, clearing such an area to turn the jungle into pastureland or farmland leads to rapid oxidation of the humus, which then becomes a hard crust. Lacking nutrients, such *laterite* is practically sterile, and it also blocks infiltration and so augments runoff. Laterite (*later* being the Latin word for brick), an excellent construction material, is often used in India and southeast Asia (in particular, the famous temples of Angkor in Cambodia), but it's bad for farming.

INFILTRATION AND UNDERGROUND WATER

Natural acid rain can eventually turn granite into clay; its effects are even more striking on calcareous sedimentary rocks, principally calcium carbonate ($CaCO_3$), laid down in the accumulated shells of ancient marine organisms. A network of caves and underground rivers riddles the Karst region (Krst in Slovenia, former Yugoslavia, not far inland from Trieste, Italy), the limestone completely dissolved by infiltrating water. Streams eat away the land to form veritable canyons, or else they simply disappear underground. Here and there, all sorts of hollows and cavities appear where collapse has occurred due to underground chemical erosion. As is evident from their varied names—*doline* (a funnel-shaped basin, from the Slovenian word), *aven* (a vertical shaft or "swallow-hole," from Celtic), or *karrenfeld* (a field marked by furrows or fissures, from German)—such features are found all over the world. In France, in the Jura Mountains to the east, and in the south, in the region of the Causses, in the spectacular sinkhole called the *gouffre de Padirac*, in the Tarn gorges, or in the "grand canyon" of the Verdon. The "Vaucluse fountain" near Avignon is the resurgence of a river that disappears underground some miles away. In southern China, tropical rains have eaten away much of the calcareous soil, leaving towers and veritable needles made of insoluble minerals, inspiring Chinese artists of the classical period. Although they appear fantastic, the drawings and paintings are in fact quite realistic renditions of these landscapes, according to those who have visited them. In the United States, enormous caverns exist under the western Appalachians—for example, Kentucky's Mammoth Cave. A particularly spectacular karstic landscape can be seen in Puerto Rico, the road taking the visitor on a roller-coaster ride from one hollow (often a hundred meters in diameter) to the next. Near Arecibo, radio astronomers have taken an especially large and regular hollow, more than 300 meters in diameter (the height of the Eiffel Tower!), and turned it into the largest radio telescope in the world. The wire mesh lining the hollow reflects the faint radio energy coming from the heavens to an antenna suspended above. In fact, the telescope collects radio waves from the zenith only, that direction moving in the sky as the Earth turns, but some adjustment of the direction of observation can be made by shifting the antenna on its cable support.

Of course, not all soils are as easily dissolved as gypsum or limestone. For other soils, the infiltration of rainwater depends on their porosity, on the activity of fauna and flora creating a labyrinth of tunnels and passageways. Even bare soil made of insoluble minerals may be permeable to water, depending on its composition. Water easily infiltrates sand or sandy soil, passing through the fine interstices between the particles, but it is soaked up by a layer of clay or blocked by some rock layers. It can nonetheless get though very narrow cracks, or seep through sandstone or other porous rocks. The pull of gravity acts on water below as well as above the ground. However, a basement of impermeable rock or soil can stop its downward flow, and the water table can even find itself trapped between two impermeable layers. The water may find a way out by piercing the surface nearly horizontally as in a spring, or else it stays trapped until freed when a hole is dug or drilled so that the water can be pulled out in a well bucket, pumped, or just allowed to flow. People have been exploiting underground water since prehistoric times, digging wells on every continent and even creating artificial oases in the Sahara Desert.[3] The pressure in the water layer corresponds to the total weight of air, water, and soil above it, and sufficiently high pressure will drive the water all the way back up where a well has been drilled. The first such *artesian* well in Europe was drilled in 1126, at Artois (France), from where the name comes. It was only much later that it was shown that much underground water is of "meteoric" origin, i.e., that it is yesteryear's rain.[4]

The atmospheric branch of the water cycle works as a still, but that does not guarantee chemical purity of the distilled water that reaches the ground as rain. Rain can contain sulfuric acid resulting from industrial pollution or volcanic eruptions, but even apart from this, it contains carbolic acid formed from dissolved atmospheric carbon dioxide. As for spring water or well water that has spent time underground, it has even less chance of being chemically pure water, inasmuch as it has been "processed" by more or less chemically active soil and rocks. Water emerging from the ground and its filtering action naturally contains minerals, sometimes more, sometimes less, sometimes salty in taste, sometimes sparkling when it escapes from high-pressure layers deep down. The varied tastes of people support the vigorous trade in mineral waters of all sorts, making it profitable to ship and fly thousands and thousands of tons of the liquid across the wide world. During the Soviet era at least, bottled water from the Caucasus flowed freely on Moscow tables. In California, you can buy bottled Evian or Perrier from France; you can also fill your canteen with an excellent sparkling water coming out of a hole in the ground next to a trail in the upper reaches of Yosemite National Park (or at least you could when I was hiking there many years ago). In some places, the ground water has circulated in calcareous formations, and it tends to be "hard." In others, with different underground natural chemistry and filtering, it comes out "soft." Some spring water stinks of sulfur (or rather of hydrogen sulfide), but after decanting, it tastes fine. Near the seashore, infiltration of saltwater raises problems, especially when the freshwater table is overexploited (a critical problem in Holland and Israel as well as in Florida).

How much water is there underground? Estimates go from 8 to 60 million cu. km,[5] an enormous range of uncertainty! In part, the problem is one of definition: to what depths do the estimates refer, and do they distinguish fresh and salty water? Soil as such contains about 66,000 cu. km of moisture. Near the surface, in the water table usually tapped by wells down to 50 meters depth, exchange of water with the surface and the atmosphere proceeds fairly rapidly, in days, weeks, or months. This includes infiltration of rainwater and snowmelt, and "exfiltration" of spring water, well water, and water seeping from underground into river and lake beds. Humans have long sought and tapped water at great depths. Thousands of years ago, the Chinese were already using bamboo shafts to drill for water down to 1,500 meters deep. In Paris, the Grenelle artesian well was drilled to 600 meters depth (twice the height of the Eiffel Tower!) in 1841, the Passy well to 641 m in 1857.

It is estimated that altogether some 3 meters (10 ft.) of water have been extracted since 1840 from the so-called Albian aquifer, lying a hundred meters under Paris. Has all this underground water been replaced by infiltrating rainwater? Or else, by upwelling water from deeper layers? And if so, how has the mineral content of the water changed? In general, the speed (or slowness) of underground water flow follows the law established by Henri Darcy in Dijon (in the Burgundy region of France) in 1856. In very compact soils (such as certain types of clay), water flows at only a few millimeters (say, a small fraction of an inch) per year, but a meter per year (1 m/y, or about 40 in/y) is quite common in more porous media. Geothermal exploitation of hot water of the Dogger aquifer, 2 km under Paris, began in 1970. The rocks go back to the Jurassic, several tens of millions of years ago, and the heat comes from the Earth's interior, but how old is the water itself? The Dogger aquifer, although deep under the Paris basin, lies close to the surface along a large arc going from the Thiérache region (renowned for its pungent cheese) near the Belgian border in northern France, to the Cotentin Peninsula of Normandy (producing excellent apple cider), by way of Lorraine, Beaujolais, and Touraine (the last two well known for their wines). Along that arc, infiltrating rainfall renews water in the aquifer, but underneath Paris, water seems to flow at only 33 cm (13 in.) per year, and isotopic analysis shows that it has been isolated from the atmosphere for at least 15,000 years and perhaps much longer.[6] One cannot estimate the rate of recharge of the aquifer on the basis of rainfall statistics for present-day climate because 15,000 years ago Europe was still in the ice age, and northern France was a cold and arid steppe.

Some underground water is much older still, its last contact with the open air going back tens or hundreds of millions of years. These ancient aquifers fascinate geologists, especially since many of them appear together with layers rich in oil or in valuable minerals. In North America, enormous pools of hot brine seem to have migrated over hundreds of miles at speeds ranging from 1 to 10 meters per year.[7] Rising from deep underground layers, the hot water, rich in dissolved minerals, underwent chemical reactions with the rock layers it penetrated, leaving behind minerals rich in lead and zinc as it cooled. Such a migrating hydrothermal system, originating under Arkansas and Oklahoma, deposited valuable minerals under Missouri to the north, while pushing oil in the direction of Colorado. Another such system may have been the source of the uranium- and oil-rich layers of western Canada. Some geologists believe that a hydrothermal (hot water) system formed under the very old Appalachian mountain range some 300 million years ago—*before* the Atlantic opened up—and left its mark not only west of the Appalachians in once-oil-rich Pennsylvania (where the wells were exhausted early in the twentieth century) but also in eastern Canada as well as in Scandinavia, Scotland, Ireland, and Greenland, now separated from North America by the spreading of the Atlantic sea floor. These migrations can only be explained by strong pressure on the trapped layers, related to motions in the Earth's crust and collisions of the continental plates, driven by sea-floor spreading.

Let's return to the surface, to the open air, and rain. Part of the water falling from the atmosphere returns to it almost immediately, evaporated from sea or land surfaces, or from leaves. What remains is divided up into runoff and infiltration (fig. 10.1). These two processes depend on the force of gravity, with infiltration important on flat terrain, runoff and consequent erosion tending to be stronger on slopes. However, a number of studies have shown that infiltration may play the major role even on hilly terrain, provided that it is wooded.[8] Following heavy rainfall, water rises quite rapidly in a stream at the foot of such slopes, without much visible erosion due to surface runoff. Instead, much of the water disappears into the ground, roots, insects, rodents, and other animals having turned the woodland soil into an extremely permeable three-dimensional labyrinth of "macropores" and even of veritable tunnels, from a fraction of an inch to as much as six feet in diameter. As a result, water easily infiltrates the soil and then flows downhill, but it also "exfiltrates" quite easily when it reaches a stream at the bottom of the slope. In this way, part of the infiltrated water quickly ends up as runoff—quickly, but not immediately. This process, a sort of underground runoff less rapid than true surface runoff, affects a layer 20 to 40 cm (a foot or two) deep. This underground detour delays and moderates the peak of a flood, provided that the soil is not already saturated by previous heavy rains. Thus although tree and plant cover will not always prevent floods, they can spread them out in time and so provide some advance warning. Such subsurface runoff plays a less important role in semiarid regions, where plant cover is spotty, the dry season is long, and even in the wet season, occasionally violent rainstorms are separated by long dry spells. This makes the soil fragile, and if it is crushed (for example, by the passage of a tractor), its permeability may be reduced by a factor as large as 5, 10, or 35.

A thick layer of permeable soil can soak up sizable amounts of water. Especially during the growing season and in the summer when high temperatures enhance evaporation, healthy plant cover will pump part of the infiltrated water and quickly return it to the atmosphere by transpiration. This process keeps the soil from becoming waterlogged; indeed it tends to dry it out. During wintertime, on the contrary, infiltration dominates, recharging the water table, especially under melting snow cover. Going to higher latitudes, approaching the pole, available sunshine decreases, the growing season becomes shorter, and snow cover lasts longer. Still, over large areas of Siberia, Canada, and Alaska, the soil and the Sun support the existence of mostly coniferous forests, the *taiga*. In many flat areas, water collects in swamps and lakes. Further north, as one reaches the *tundra*, trees disappear from the landscape, leaving only berries and flowers (as well as grass) growing in a brief summer season. Over much of the subpolar zone, the *permafrost* soil, frozen since the last ice age, still remains frozen for at least part of the year. In all these regions, winter is followed by a season of wretched roads, mud, and mosquitoes.

Infiltration involves not only the flow of water but also chemistry. Chemical properties of the water reaching surface streams or the underground water table depend both on the acidity of the falling rain and on its interactions with the plant cover, the humus, and other constituents of the soil. Specialists of *hydrochemistry* seek to follow closely the transformations of pollutants in wooded watersheds. In a number of areas, in Europe and in North America, they keep track day by day, sometimes hour by hour, of the water entering and leaving a small watershed, along with weather changes.[9] On a few small sites (catchments of order 10 hectares or 25 acres), they carry out true experimentation, as opposed to passive observation, deliberately introducing nitrates or sulfates with an isotopic "fingerprint" that can be followed, and monitoring the reaction in the experimental basin. Such experiments have shown that dry deposit, from the atmosphere, of sulfur and nitrogen in the form of ammonium sulfate can have dramatic immediate effects on the composition of the water flowing out of the basin as well as on leaching of the soil. The aim of such research is to check whether computer models of the hydrochemistry of wooded watersheds can indeed be trusted to simulate the response of these complex natural systems to anthropogenic chemical perturbations such as acid rain and fertilizer runoff.

Runoff: From Rills to Rivers

Let's look more closely at running water on the land. Wherever water falls on a sloping surface, it tends to flow downward under the pull of gravity, and this is what runoff is all about. On plant cover, water drips along and from leaves and branches, along stems and tree trunks. On the ground, the water not immediately soaked up by the soil flows along a network of changing rills and gullies, finally entering well-established brooks and streams. Not many raindrops reach the ground directly in a thick forest. However, on fields and farmland, and also on sparsely covered semiarid woodland, large raindrops can change the soil's properties by their impact, shattering clumps of dirt and scattering dust particles. That doesn't make much difference for land plowed by the farmer, or tunneled by moles, worms, ants, and termites. However, on bare soil, especially in semiarid regions where when it rains it pours, the dust scattered by the raindrops can stop up the pores of the soil, forming a sort of impermeable crust. In arid regions, in the deserts of the southwest United States and elsewhere, the rare but violent cloudbursts give rise to extremely dangerous flash floods. In any case, whenever water reaching the surface begins to accumulate, whether because the soil is saturated or its permeability too small, or the rainfall rate is too high, the water will flow. Sometimes, for nearly flat very little inclined surfaces, the water flow spreads out and carries away the dust particles scattered by the raindrop impact: an unspectacular so-called *sheet* erosion, nevertheless capable of moving large amounts of soil and shaping the landscape. Most of the time, however, the flow of water finds a path of least resistance along the steepest slope, forming an intricate network of gullies. In arid and semiarid regions, that produces ravines with a variety of names: *arroyos* (Mexico and the southwest United States), *oueds* (North

Africa), *wadis* (Near East), etc. Erosion also acts in more or less humid climates, especially as the flowing water carries along all sorts of soil and plant debris. On slopes without contour plowing or terraces, the flowing water, and the abrasive particles that it carries with it, eat away the soil to form ditches and even veritable ravines. For sandy or clayey soils, these have a V-shape, for others (loamy) a U-shape, and in either case they can be several meters deep. Once such ravine formation has started, it tends to maintain itself because it creates steeper slopes down which the water flows faster. Except in some hyperarid regions where the surface water either disappears underground or evaporates completely, such ravines end up supplying brooks and other more substantial streams with water already heavily loaded with abrasive particles.

What about snow? On the plains, water falling in this solid though apparently insubstantial state accumulates throughout the winter. Water from melting snow will infiltrate and saturate the soil, recharging the water table. Throughout the period of cold, transpiration and evaporation slow down or come to a halt. Sublimation—the direct passage from the solid to the gaseous state—may occur, and it becomes significant when a dry warm wind blows. Such winds, called *foehn* in the Alps, *Chinook* in North America, can vaporize a thick layer of snow in a matter of hours, to skiers' dismay. In the circumpolar zones and on high mountaintops, where temperatures rarely if ever rise above freezing, each year's snowfall piles up on the previous years.' Its properties change as it lies exposed to the air, to changing temperatures, and to the increasing weight of layers of fresh snow that fall on top of it. It's often written that the peoples of the north (Inuit, Yupik) have hundreds of different names for snow, and all around the Arctic many constructions and composite words are needed to describe its varying qualities; after all, even skiers from the city use more than the single word *snow*.[10] Glaciologists use the Swiss-German word *firn*, last year's snow, to designate one of the transition phases between freshly fallen (powdery) snow and the ice found deep in glaciers or the great ice caps (the latter called *inlandsis* by specialists, using the Danish word). Thus water in its solid state accumulates in mountain glaciers and the inlandsis of Greenland and Antarctica; and the glaciers, although solid ice, flow downward under their own weight, sometimes slipping on a layer of meltwater when temperatures are higher at the bottom than at the top of the ice sheet. The top layer of snow, out in the air for a time, traps air bubbles and collects wind-borne dust from distant deserts, as well as ash and other volcano eruption ejecta, particles of lead and other products of mining and industrial activities, and radioactive fallout from nuclear explosions and accidents. Each year's layer is covered by the next years' snow, and so the record of what has been in the air is kept on ice until scientists find the resources, energy, and tools necessary to extract it. Advancing mountain glaciers gouge the rock and carve out valleys, dragging debris along. After they recede, as happened only recently (8,000 to 15,000 years ago) in North America and Europe, their mark remains on the landscape: *cirques* and valleys of different shapes, chains of lakes in Switzerland, Scandinavia, Canada, and the United States (in particular, the Finger Lakes of upstate New York). Marking their farthest advance, the glaciers left behind the *moraines* (i.e., deposits of sand and other alluviums), as in the case of Cape Cod and Long Island, or in Europe the hills of Pomerania (Poland and northeast Germany).

In Antarctica, Greenland, Spitsbergen, and Alaska, glaciers flow and crash into the sea, breaking up, sometimes into giant icebergs, sometimes into veritable ice islands many miles across, sometimes shattering noisily and spectacularly into smaller ice fragments (nonetheless treated with respect by the masters of Alaskan excursion boats). This completes the long cycle: once evaporated long long ago from the sea and fallen some time afterward as snow from the atmosphere, the water finally returns to the sea after spending tens or hundreds of thousands of year locked up as ice. Water generally spends less time in mountain glaciers, although in some cases its frozen sojourn lasts tens of thousands of years, and it seldom returns directly to the sea. Instead, meltwater of mountain glaciers flows mostly into the waters of the land, into torrents, streams, and finally rivers (fig. 10.3). In Switzerland you can drive along the Rhone River up to the foot of the glacier that is its source, and then, a few miles from there on the other side of the Grimsel Pass, you can visit the lakes that feed the Aare, a river that flows into the Rhine just upstream of Basel (in Switzerland, at the triple border with France and Germany). Other tributaries of the Rhine have their source a bit further to the east. A few miles' difference in where snow falls in Switzerland determines whether the water will end up in the Rhone and the Mediterranean or in the Rhine and the North Sea. Further north, separating the Black Forest in Germany and Alsace in France, the Rhine receives water flowing off the slopes of the Kandel, a small 1,241-meter (4,068 ft.) summit northeast of the university town of Freiburg. Only a few miles from there is the *Donau-Quelle*, the source of the Danube River,

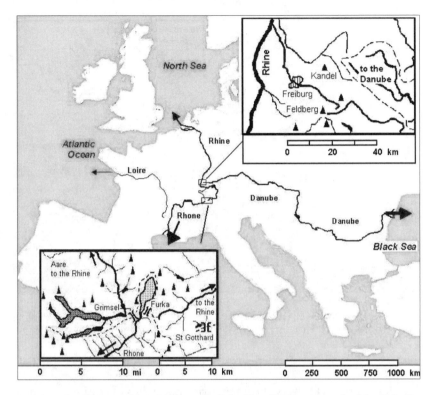

FIGURE 10.3 European rivers. In France the Loire River flowing west to the Atlantic begins only a short distance from the Rhone River. The separate destinies of Alpine snowfall appear in the inset map of the vicinity of Furkapass (Switzerland), where both the Rhine and Rhone Rivers originate. Summits above 3,000 meters are shown. The upper right inset (summits in the range 1,000 to 1,500 m) shows part of the Black Forest (Germany), where very close to the northward-flowing Rhine, the Danube begins its long journey east to the Black Sea by way of Vienna, Budapest, and Belgrade.

which flows east to empty into the Black Sea. In Asia, precipitation on a single much larger mountain massif supplies the water of several of the world's greatest rivers (fig. 10.4). The Brahmaputra, Irrawaddy, Chang Jiang (Yangtze Kiang), and Huang He (Yellow River) all originate on the slopes of the Himalayas or on the Tibetan plateau, where moisture-laden Indian monsoon air, forced up by the topography, dumps the highest rainfall in the world. And several hundred miles to the west, such rain also feeds the Indus and Ganges Rivers, the one flowing to the Arabian Sea, the other to the Bay of Bengal on the opposite side of India.

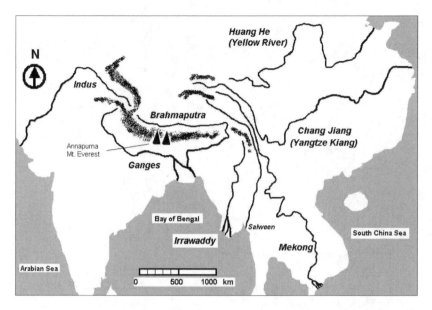

FIGURE 10.4 Asian rivers. Separate destinies of rain and snow on and around the Himalayas, the Pamir and Hindu Kush further west, and the mountains of Szechwan and Yunnan to the east.

From brooks to rivers, the water carries the debris of erosion, and these debris act in turn as erosive agents. The smallest particles (diameter less than 50 microns, say 0.012 inch) are easily moved along with the flow, but they accumulate as silt if the current is weak. When the river overflows its banks, they are left behind as loamy alluvia, collections of the extremely fine particles that result from smashing up all sorts of material—plant debris, gravel, clay, pebbles, and sand—carried along with the water in its descent with the mountain torrents. Where the mountain stream or glacial outflow reaches the plain, the deposits spread out in an alluvial fan. But if the climate dries out, such loamy soil is easily blown away, dust in the wind finally settling down elsewhere as *loess*. In northern France, Belgium, and Germany, such loess deposits bear witness to the arid conditions that prevailed over parts of Europe not covered by ice during the glacial advances of the last two million years. Loess layers dozens of meters thick cover much of the Yellow River (Huang He) basin of China, with up to 200 meters of loess on the plateau above. These extremely fertile soils are very vulnerable to erosion by water or wind. Flowing water carves out deep ravines and valleys, leaving spectacular pinnacles or natural bridges where less crumbly soil resists erosion. The Yellow River's color comes precisely from the huge amount of loam that it carries, more than a billion tons every year, 12 percent of the world's river sediment load (against only 0.1% of the world's river flow). Fortunately, not all of this good earth is carried off to the sea, part being deposited downstream of the loess plateau, raising high the banks on each side of the river, veritable levees.[11] During dry periods, dust blown off the loess plateaus falls on Beijing, and sometimes thousands of miles further east as far as Alaska.

River flow varies through the year, depending on the melting of upstream snow as well as on rainfall. After abundant snowfall during winter and spring, mountain torrents flow furiously, displacing enormous boulders, uprooting trees along their banks, stirring up stones and gravel, carrying off some of this debris to the plain. Still, inasmuch as snowmelt is a gradual process, such floods are not the most dangerous, and as the summer season advances, stream flow dwindles. Much more dangerous floods result when violent storms dump large amounts of water (10 in. of rain amounting to much more than the same thickness of snow) in a matter of hours. People hiking, fishing, cycling, or driving in a narrow mountain valley should certainly not ignore the risk of rapidly rising water. Even in the plain, flash floods can cause enormous damage where a stream emerges at the foot of the mountains. The town of Fort Collins, home of Colorado State University, was badly hurt by the flash floods of the Big Thompson River in 1976 (145 dead) and again in 1997.[12] Earlier, on June 10, 1972, the Cheyenne River emerging from the Black Hills wreaked havoc on Rapid City, South Dakota. That calamity was the consequence of an extremely violent long-lived storm, whose heavy rain saturated the soil of the slopes and caused a mudslide into the already swollen stream. Similar floods hit the Languedoc region of southern France from time to time, fortunately mostly with less loss of life. The Sun shines much of the time, and rain is rare; but heavy rain can fall, and when it does it sends floods down the usually dry arroyos, locally called *cadereaux*. Thus, on October 3, 1988, 420 mm (more than 16 in.) of rain fell in less than seven hours on the hills surrounding Nîmes, and all that water almost immediately flowed through the town center. Of course, it must be admitted that the scale of that disaster, and of others like it, is often magnified by all the paving and other "improvements" which reduce the permeability of the area, and by the fact that many new buildings are put up in harm's way (as they were in Nîmes, next to the *cadereaux* where the water had to flow).

In other arid regions, purely natural factors account for the damage caused by floods. During El Niño events (see chapter 7), it can rain in the desert at the foot of the Andes along the Pacific coast from Peru to northern Chile. Although a few rivers fed by snowmelt in the Andes traverse this area, the land is usually bare with hardly any vegetation. When rain falls, it rapidly saturates the soil, and with little loss by evaporation and none by transpiration, it accumulates at the surface and flows in the direction of the Pacific. This runoff transforms the normally dry ravines called *quebradas* into running streams and swells the trickle of the rivers into a formidable flow. The extremely strong 1997–98 event was only the last of the major El Niño events of the twentieth century, which include 1982–83 and 1972 as well as others that took place before the era of space observation. For example, in March 1925, after three days of more than 200 mm (8 in.) of rain, the Rio Moche (which empties into the Pacific about 300 miles north of Lima, Peru) rose by more than ten feet, washing out roads, bridges, and devastating the town of Trujillo with severe loss of life. Two centuries before, on March 15, 1720, a similar flood destroyed the village of Zaña (also in northern Peru), leaving its high-water mark on the walls of the only building that survived, a convent. During the 1972 El Niño, which no one had predicted, a team of archaeologists from Chicago's Field Museum, who had come to study the pre-Columbian civilizations of this region, discovered the traces of a still greater catastrophe.[13] Agriculture had been practiced in this area since 1000 B.C., and the inhabitants had dug irrigation canals to exploit the water of the Rio Moche and extend their farms to the *quebrada rio Seco*. To the north, they built the village Cerro la Virgen, and they created a civilization that reached its height in the ninth century, with a capital city (Huaca del Sol) and a distinctive (Chimu) style of pottery. It seems that an exceptional El Niño, toward the year 1100 A.D., caused a catastrophic flood that sent this civilization into decline. The Chimu did build a new capital after the flood, but it was taken by the Incas about 1465; the Spanish conquistadors came on the scene only in 1532. In their diggings between the sites of Huaca del Sol and Huaca de la Luna, the Field Museum archaeologists found ancient bricks that had lost much of their shape as a result of being saturated with water, as well as damaged irrigation canal works. They concluded that the flood stage must have reached as much as 15 meters (50 ft.) above normal river level, with widespread erosion of much of the alluvial plain. The damage wreaked by the El Niño rains of circa 1100 was aggravated by the fact that tectonic movements (quite common between the Andes and the Pacific) had raised the area by several meters above its level of 500 B.C., steepening the slope of the terrain and intensifying effects of erosion. Such shifts also seem to have played an important role in the dramatic changes in the course of the Yellow River (Huang He) in China. Before 1851, only 150 years ago, it emptied into the Yellow Sea (Hoang Hai). But seismic activity led to a change of course, and since then it empties into the Gulf of Chihli (Bo Hai), a few hundred miles to the north.

ALONG THE GREAT RIVERS

Each of the world's great rivers has its own path and its own story. Very little separates the sources of the Rhine, the Rhone, and the Danube, and the three rivers all stay in the same temperate climate zone, but they are far from identical. Some similarities exist between the Danube and Rhone deltas, but hardly with the Rhine in the Netherlands. What about other continents? Both the Nile and the Congo Rivers get much of their water from central Africa, but what a difference in where they go! Starting from Lake Tanganyika and the Lukuga and Lualaba Rivers, the Congo River flows through the tropical rain forest draining a vast watershed of 3,684,000 square kilometers (1,422,000 sq. mi.) where on the average 900 to 3,000 mm (35 to 120 in.) of rain fall each year. Swollen by the contributions of its many tributaries, it carries a couple of trillion tons of water to the Atlantic at an average rate of 50,000 tons per second. The Nile also starts from central Africa, but it flows toward the arid zone in the north. Actually, the very long White Nile starting at Lake Victoria (shared by Uganda, Kenya, and Tanzania) has relatively modest flow. The tumultuous Blue Nile, joining the White Nile at Khartoum (Sudan), brings most of the water, at least during the six months of the year when it is swollen by rains that fall on the hills and mountains of Ethiopia. Over most of its course, more than 3,000 km (1,800 mi.), the Nile receives no water at all; on the contrary, it continuously loses water by evaporation as well as by evapotranspiration of the crops grown since thousands of years by Egyptian *fellahs*. Studying satellite photos, or looking out the window when flying over Egypt, you can see that vegetated land constitutes but a narrow green band limited east and west by the vast expanse of desert.

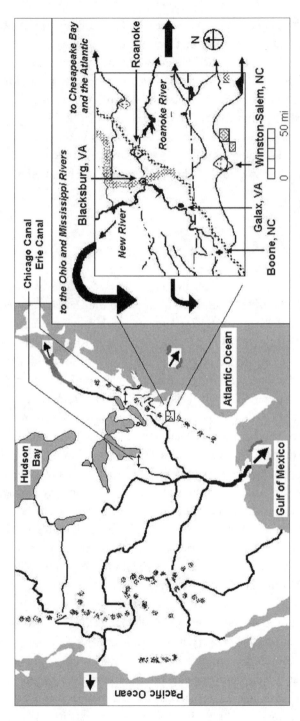

FIGURE 10.5 Water in North America. The Continental Divide (between waters going to the Pacific and the Atlantic) follows the Rocky Mountain chain in the western United States and Canada. Only low hills divide waters going to the Atlantic by way of the Great Lakes and the St. Lawrence River, and waters going by way of the Mississippi and the Gulf of Mexico. Economically important links exist through the Erie Canal, providing a shortcut from the Great Lakes to the Atlantic by way of the Hudson River and New York, and through Chicago to the Illinois River and the Mississippi. As shown in the inset, the Blue Ridge Mountains of Virginia and North Carolina separate waters going directly to the Atlantic by way of the James, Shenandoah, Potomac, and Susquehanna Rivers, among many, from waters going by way of the Ohio and Tennessee Rivers to the Mississippi.

Other striking contrasts appear between the Brahmaputra and the Huang He, although both are born in the mountains and highlands surrounding the Tibetan plateau. Just like the Congo, the Brahmaputra reaches the sea after flowing through well-watered tropical land, in this case Bangladesh.[14] As for the Huang He, flowing north it enters the arid interior of China, with less than 300 mm (12 in.) of rain a year. Like the Nile, it creates a green band where it crosses the desert of Inner Mongolia. Only where it turns back south, flowing again to the south of the Great Wall, does the Huang He enter a zone where some rain falls; with additional water from its tributaries, the Yellow River then supports intensive agriculture and very high human population density.

In the Americas, the fate of water from the skies also depends on just where it falls, on one side or the other of the continental divide (fig. 10.5). In the northern Rocky Mountains, water ends up either in the Columbia River's flow to the Pacific, or in the Mississippi-Missouri system flowing to the Gulf of Mexico. In the state of Colorado, snowmelt feeds on the one hand the rivers flowing to the east and the Missouri, on the other the Colorado River heading southwest through the desert, through the Grand Canyon that it carved out, to the Gulf of California. In the east, Appalachian rainfall flows either toward the Mississippi or eastward on the short direct path to the Atlantic. In South America, the separation between the Orinoco and the Amazon River basins is not very clear, and indeed the two systems are connected in the Amazonas province of Venezuela. In Brazil, on the plateau of Mato Grosso, some of the water flows to the Amazon, some to the Paraná-La Plata system. Still, whether in North or South America, the dominant phenomenon seems to be the continental-scale *collection* of water coming from many sources fed by rain falling over extended areas. The Amazon River's flow, 200,000 cubic meters per *second*, is the largest in the world. It results from collection of water descending from the Andes over a distance of more than 2,000 km (1,200 mi.) from southern Colombia to the *Altiplano* of Bolivia, and from drainage of an enormous watershed, more than 6 million square kilometers (2.3 million sq. mi., roughly nine times the area of Texas), moreover a very well-watered basin with 1,500 to 3,000 mm (60 to 120 in.) of rain per year. The river's slope is extremely small over most of its course, and sizable ships make the 4,000-km trip from the Atlantic all the way up the river to Iquitos, Peru, at the foot of the Andes. Continual exchange of water between the atmosphere and the dense vegetation prevails over most of the basin, and so a large fraction of the rain falling in the interior consists of water recycled by evapotranspiration, although the moisture, and even some snow falling on the Andes, "originally" comes from evaporation of Atlantic Ocean water.

In North America, the Mississippi-Missouri system collects water from three principal sources spread out over a watershed of more than 3 million square kilometers (over 1.1 million sq. mi.) and delivers 500 billion cubic meters per year to the Gulf of Mexico (fig. 10.5). The longest branch, the Missouri River, collects a large part of the snowmelt on the eastern slopes of the Rockies from Montana to Colorado, the snowfall a consequence of the uplift of humid air masses mostly coming from the Gulf of Mexico. After crossing relatively dry Montana, the Missouri picks up the Yellowstone and Platte River flows, also coming from the Rockies, and flows on through progressively less dry areas of the Great Plains, finally joining the Mississippi just above St. Louis. The sources of the Mississippi itself lie in northern Minnesota, where the river drains a fairly well-watered area. At Hannibal (made famous by Mark Twain), 120 miles north of its junction with the Missouri, the Mississippi is already impressively wide. Downstream of St. Louis, at Cairo, Illinois, it takes up the abundant flow of the Ohio River, draining a very well-watered area including in particular the western slope of the Appalachians from Pennsylvania to Alabama. Further south, near Memphis, the Mississippi flows by a beautiful campsite, where you can be eaten alive by mosquitoes in the wrong season. On the last stretch to the Gulf of Mexico, the Mississippi picks up still more water from the Arkansas River coming from the Ozark Mountains. In Louisiana it is joined by the Red River, well known to movie-lovers both under its own name and because it flows near Paris, Texas (also the title of a Wim Wenders film perhaps better known to Europeans than to Americans). The Mississippi carries enormous amounts of sediments with it, depositing alluvia all along its meandering course, forming natural dikes or levees as well as the delta before sending the rest into the Gulf of Mexico. The Mississippi, like other great rivers of the temperate zone, flows in its "minor" bed, but from time to time its waters rise and overflow its banks, flooding a much larger area, the "major" bed or floodplain. The deposits of silt and mud make the soil of the floodplain extremely rich, but building there is asking for trouble. Muddy floodwater fertilizes fields, but it blocks roads and soils houses. Every so often, flooding occurs on a spectacular scale, despite (or because of) all the efforts of the Corps of Engineers to control the waters. In 1993 the city of St. Louis was nearly surrounded by floodwaters of the Mississippi and Missouri. Within the floodplain, the Mississippi (and on much smaller scales, the Seine River downstream of Paris, the Jordan south of the Sea of Galilee) follows a meandering course. With alluvial deposits on one bank and erosion on the other, the riverbed tends to shift, finding new meanders and abandoning old ones sometimes together with chains of ponds. Sometimes that leads to embarrassment when the riverbed has run dry just where a bridge was built, with the river gone to flood the road or railway elsewhere. Other rivers make their bed in several more or less straight, more or less shallow channels, within a broader bed of sand, gravel, and sometimes small islets.

Floods are usually counted as natural disasters, but catastrophic damage and loss of life are in part the consequence of careless land use and construction in river floodplains. Floods also fertilize farmland. When the river rises and overflows its banks, the spreading floodwaters slow down or stagnate, leaving time for suspended solids to settle out. These alluvial deposits are so much eroded topsoil from upstream saved for the land. Of course, the Mississippi ("Big Muddy") hasn't lost all its load of sediment at its entry in the state of Louisiana. In the last hundred miles, suspended solids help to form and continually reconstitute the Mississippi delta land and bayous of the Cajuns. In China, alluvial deposits due to flooding by the Huang He and the Chang Jiang have played a key role in maintaining fertile farmland, but precisely because this has made high population density possible, it has increased the population at risk from floods. Over the centuries, floods in China have claimed hundreds of thousands, indeed millions of lives, but war and civil strife have often contributed to these catastrophes.

Several great rivers flow in northern forests, crossing tundra and permafrost to empty into the Arctic Ocean: Canada's Mackenzie River, the Ob, Yenisey, and Lena Rivers of Siberia (figs. 10.4 and 10.5). Other important rivers flow into salty inland seas that lose their water by evaporation: the Volga and the Ural Rivers into the Caspian Sea, the Amu Darya and the Syr Darya into the Aral Sea (fig. 10.2). In fact, the Aral Sea has been shrinking for many years and has now split up into several shallow saltwater lakes, partly because of gross overexploitation of its water for irrigation of cotton fields in the former Soviet republics of Central Asia. It may even dry out completely in coming years. However, the level of the much bigger Caspian

Sea has fluctuated from one decade to another, and some geographers note correlation between the level of the two seas. The two bodies of saltwater are separated by the terrible Ust-Urt Desert, but could they be connected by a deep underground water table, and if so, why should there be a see-saw of water level between them?

Another river, not very prepossessing physically, emptying its feeble flow into another inland sea, occupies nonetheless a most important place in the minds of the world's people (or at least in those of the Jewish-Christian-Muslim tradition): the Jordan River, ending in the Dead Sea. Fed by snow melting on the slopes of Mount Hermon and by other streams coming from springs in Israel, Syria, and Lebanon, the Jordan first flows into the Sea of Galilee (called by some the Lake of Tiberias, but Kinneret in Hebrew according to its lyre shape). A living sea of fish, the Sea of Galilee is a freshwater lake which remains fresh because fresh water flowing in from upper Jordan replaces water flowing out to the southern lower Jordan River or pumped out for Israel's National Water Carrier, supplying central and southern Israel. But from there on, the Jordan descends into a deep valley, in fact a rift in the Earth's crust that is the extension of the *East African Rift*. Less than a hundred kilometers (60 mi.) south of the Sea of Galilee, the river's waters plunge into a depression that they only half fill. At 396 meters (1,300 ft.) below sea level, the Dead Sea is cut off from the Mediterranean by the Judaean Hills, and from the Gulf of Elath/Aqaba and the Red Sea by an arm of desert that extends from the Sahara by way of the Israeli Negev and the Kingdom of Jordan to Arabia. In this dead end, Jordan waters can exit only by evaporation, leaving accumulated salt behind. Loaded with minerals and strongly salty, these waters support practically no life, apart from some remarkable bacteria—a Dead Sea indeed.

WATER ON THE CONTINENTAL SCALE

As great collectors of water, essential links in the continental part of the water cycle, rivers have always played a major role in the development of civilization. And considering the number of people on this planet today, human well-being depends more than ever on the availability of water. But today's human activities include not only water withdrawals but also large-scale water pollution. And we still have much to learn about what happens to the water that falls on the land. In many countries, especially rich ones, hydrologists, agronomists, hydrogeologists, and other specialists are hard at work trying to advance understanding of infiltration, runoff, and mechanical and chemical erosion processes under varying conditions. Much progress has been made, in particular in modeling local- and regional-scale processes, thanks to laboratory and field experiments that make it possible to check and improve the models, which can then be applied to practical problems. However, in recent years, it has become clear that scientists need to extend the field of investigation, to look beyond the fascinating but limited problems of small watersheds or catchments that are less than a hundred miles across.

The organization of larger-scale field studies has indeed begun in different parts of the world. In southwestern France (not very large-scale for Americans), the cost of the severe drought in 1989–90[15] incited the Météo-France research center to organize, together with other scientists, a pilot "experiment" on atmospheric-hydrologic interactions, so as to better model the soil's water balance. Although called an experiment, this was really an intensive campaign of field observations, with instruments set up in a dense network covering an area of about 400 square miles to get both meteorological and hydrological data, and with analysis of complementary data from weather satellites, research aircraft, and from more than the usual number of weather balloon ascents. Following this successful campaign, another one was organized in semiarid West Africa. There have of course been many similar campaigns in the United States, in particular around the ARM (not military, but Atmospheric Radiation Measurement), southern Great Plains site of the U.S. Department of Energy, in Oklahoma. Such "ground truth" helps to understand and check what scientists call remote-sensing data, in particular data from weather (NOAA, the National Oceanic and Atmospheric Administration, etc.) and Earth resources (Landsat, SPOT, etc.) satellites.

In recent years, more and more research establishments and scientists have taken part in studies of hydrological processes on the continental scale, nearly always necessarily involving international cooperation. For the enormous Mississippi-Missouri basin, new comprehensive data obtained on the ground, in the rivers, in the air, and from satellites are available to research scientists of many countries, not only the United States. Similarly, Canada runs the Mackenzie basin study, but without reserving the data for Canadian scientists only. For the biggest river basin of the world, Amazonia, Brazil co-ordinates research on how water, heat, and carbon are exchanged between the air and the surface water, vegetation, and the soil. In Asia, the winter and summer monsoons dominate precipitation in southeast Asia on the one hand, Tibet on the other, with the water flowing through several different river basins, part south, part east, and part north up to the mouth of the Lena River in the Arctic Ocean. Japan coordinates these studies with strong par-ticipation by China and Russia. In Europe, more the western peninsula of Eurasia than a true continent, a large-scale study focuses on the Baltic Sea, which can be considered part of a watershed (fig. 10.3). The Baltic, a nearly closed shallow sea, partly covered by ice during the winter, receives fresh water, heavily polluted by urban, industrial, agricultural, and mining wastes, from many rivers from the surrounding countries. Baltic water reaches the North Sea by way of the narrow Øresund strait between Denmark and Swe-den, flowing at a rate of 470 billion cubic meters per year, comparable to the Mississippi's outflow. A dozen countries of eastern and northern Europe take part in the BALTEX program, coordinated at the Geesthacht (Germany) re-search center near Hamburg.

There is always room for new ideas in research. The orbits of artificial satellites depend on the distribution of mass in the Earth. Today, more than forty years after the start of the space age, scientists have distilled the fine de-tails of the Earth's gravitational field and exactly how these influence satel-lite orbits. At least 70 trillion tons of water exist in the form of soil moisture, but little is known about how this is distributed among different watersheds, and how it shifts and changes from one season to the next and from year to year. Even more water lies deeper underground, but evaporation removes 70 trillion tons from land surfaces every year, about the same amount of water as exists as soil moisture. Atmospheric humidity and precipitation depend in part on this supply coming from the poorly known stock in the soil (fig. 6.1, table 6.1). Soil moisture in a particular place can be measured using a neutron sonde, but this varies enormously, even from one cornfield to the next. With the joint NASA—DFVLR (German Air and Space Research Es-tablishment) Gravity Recovery and Climate Experiment (GRACE), scien-tists from Austin (Texas) and Potsdam (Germany) hope to be able to follow shifting masses of water in the soil and the oceans by way of their minute ef-fects on the satellites' orbits.[16] It's quite a challenge, but since 1992 the NASA—CNES (French Space Agency) Topex-Poseidon satellite has been measuring sea level over the planet with an accuracy of a centimeter or two (less than an inch), and following ground water is worth trying.

Apart from the thirst for knowledge, are there other good reasons for scientists to devote their professional lives and energy on such research, also spending taxpayers' dollars, pounds, yen, or euros? As already noted here, and I'll come back to it later, concern about climate change is in the air. Indeed, humankind is engaged in a global "experiment," perturbing the planet's natural "biogeochemical" cycles, in particular by emissions of carbon dioxide. But as an experiment, it's sloppy: there's no "control"; in other words, there's no parallel unperturbed planet whose climate history we can compare with that of our own Earth. If we don't want to just wait to see how the experiment turns out (with us in the test tube!), if we want to have some idea of the way in which climate is likely to change under the influence of our own actions, our only option (apart from looking into crystal balls or consulting cards or tea leaves) is to use the tool we have: modeling. For nearly five decades now, scientists have developed climate and weather prediction "models." In such models, they seek to represent the complexity of the real climate system—air, clouds, rain, snow, oceans, ice, fresh water, and land—using a few equations based on well-established laws of nature, and applying them to a limited number of quantities, *parameters*, describing the state of the system. Since the 1970s, the favored approach has been to put the equations into numerical form and to use powerful electronic computers, introducing measured (or interpolated) values of the parameters at a given instant of time, and solving the equations to determine the future values of those parameters. Thus one "runs" the "general circulation models" or GCMs that are presumed to represent, at least schematically and on a large scale, what goes on in the atmosphere and ocean of the planet. With such models, numerical experimentation is possible: just change the input data according to one or another scenario (for example, increasing emissions of carbon dioxide), and the model computes a possible climate future. How believable are such projections? To have confidence in them, one needs first to show that the models do a good job representing what goes on in today's climate. Although the models do quite well in many respects, plenty of problems remain. For example, a certain model may give correct precipitation over Arkansas in the summer but not in the winter. But if in a particular model, computed winter rain or snow is too small, it won't recharge soil moisture adequately and the (computer-world) land will dry out in the summer, crops and plants generally will transpire less (sending less water back to the atmosphere), and summertime model rain should decline. That may not make too much difference for weather forecasts, even several days in advance, because the weather prediction services introduce (the technical word is "assimilate") new observed data every twelve hours, preventing the model from straying too far from reality. But for long-term climate simulation, the same model will do a poor job, and one may wonder about its reliability in predicting a real-world climate change in response to a given perturbation. Reducing uncertainty requires that bad models be thrown out; and what better method than testing them ever more severely with finer and finer measurements over the entire globe? One fundamental principle guides such analyses: water and energy always flow and change their form on our planet, but they never vanish, nor are they created.

PART III

Water in Human History, Past and Future

CHAPTER 11

WATER AND MAN'S RISE TO CIVILIZATION

OUT OF THE ICE AGE

Everyone talks about the balance of nature, but hasn't it been overrated? Think of all the upsets over the last few million years since our ancestors distinguished themselves from chimpanzees (or vice versa), long before hominids had any significant impact on the environment! Our ancestor Cro-Magnon Man (and Woman!), like other plant and animal species, had either to migrate or to adapt to the repeated advances and retreats of the ice caps and of sea level. Conditions comparable to those that we enjoy today, with extended ice caps limited to Antarctica and Greenland, reigned during perhaps twenty different intervals over the past few million years, but no such interglacial interlude has ever lasted much longer than 12,000 years. The rest of the time, in fact most of the time over the last two million years, ice sheets covered up to three times as much land as today. Extended forests of leafy trees covered much of Europe, eastern Canada, and all of New England a little over a hundred thousand years ago. But those forests disappeared well before anyone took an axe to them. Only 18,000 years ago, practically the day before yesterday, much of Europe downstream of the ice sheets had no trees, only permafrost, tundra, a semiarid steppe, and prairie. Five thousand years ago, yesterday so to speak, dense forest had returned to many of those regions; today corn, wheat, and potato fields have replaced some of that forest.

During glacial times when ice covered eastern Canada and northern Europe, the great deserts expanded and stronger winds blew poleward, leaving abundant deposits of desert dust in the layers of ice corresponding to those glacial advances. In today's West African Sahel, the transition zone between today's Sahara Desert to the north and savanna land to the south, fossil sand dunes of the more extended desert of the past can still be recognized. Although the zone around latitude 15° to 20° North has indeed dried out over the past 3,500 years, following a warm moist period some 5,000 years ago, vegetation there is still much more abundant than it was during glacial maximum. Today, acacias tend to grow in the hollows between the fossil dunes, while other trees and bushes grow on their crests. The great change between glacial and interglacial periods necessarily involved changes in atmospheric circulation, modifying precipitation and evaporation (i.e., the land's supply and loss of water), so that the climate zones and the borders between the Sahel, the savanna, and the dense tropical forest (jungle) also had to shift.

In parallel with such natural changes in the environment, the impact of human activities has certainly grown. Still, can we speak of upsetting the balance of nature, when for at least two million years there has been no balance? Once only predators somewhat more successful than others, humans with their new activities began to put their mark on the environment—on a small scale at first but gradually more and more, accompanying, reinforcing, sometimes offsetting the vast transformations under way with melting of the ice, erosion of new valleys, and advancing forests and vegetation reconquering the north and encroaching on the Sahara. Even today, the "natural" environment continues to evolve on its own and not only in response to human pressures. The memory of the ice age still marks the Earth's crust as well as the course of rivers. In many places, the soil, the landscape, and the areas populated by different flora and fauna still evolve today in response to the less frigid conditions that set in only recently, about 10,000 years ago.[1] *Panta rhei!* Nature is in a state of change, not of balance.

Picking and eating fruit that they found tastiest, tossing the pits by the path, men and women set into motion and accelerated the processes of natural selection and genetic modification, without at first knowing just what they were doing.[2] Such interference in natural selection has not always worked out well: it seems that by excessively successful hunting, the first inhabitants of the Americas drove Western Hemisphere species of great mammals to extinction. On many occasions, the Polynesian colonization of previously uninhabited Pacific islands seems to have led to outright eco-logical catastrophe.[3] More and more, humans have become the principal actors on the stage of planetary change. At first, they could only submit to the vagaries of nature, while trying, often successfully, to take advantage of change. Or is it that only those who succeeded in taking advantage of the changes left any descendants? Some time between 10,000 and 8,000 years ago, when the ice had gone, apparently for good, the new warmer and rel-atively stable climate finally rewarded those who had the idea of planting and growing crops rather than going off hunting and collecting food. Do-mesticated barley and wheat have been found at Jericho in carbon-14 dated pottery shards 9,500 years old. The "invention" of agriculture fundamen-tally changed human existence, with profound consequences for the future of the biosphere as a whole. But agriculture only makes sense if one can count on water. Without a dependable supply of water, there's no point in putting down roots, and in semiarid regions where one can only hope for rain without knowing in advance exactly where it will fall, nomadic peoples continue to resist the idea of settling down. Better to live there as nomads, driving the herds to the green rain-fed pastures of the season.

Agriculture requires a reasonably reliable source of water, even if it cannot be fully controlled. Where water flows freely, people can be tempt-ed to become sedentary, sow crops, and take root. Farming is attractive even in a land of desert or semidesert clime, provided that a river coming from well-watered regions runs through it, occasionally overflowing its banks to lay down its load of silt. So long as the river continues to flow, farmers can water and irrigate their crops, taking water directly from the river or from wells dug in the nearby water table; and they can dig ditches to bring the water directly to the fields. From time immemorial—*no!*— since several thousand years only, humans have used the waters of the great rivers—the Nile, the Tigris and Euphrates, the great rivers of the Indian subcontinent and of China.

Such exploitation of river water, essential in dry regions, has its advan-tages even in those areas that enjoy fairly abundant rainfall. After all, rains vary, and for some crops it's better to water a bit every day and not wait for downpours that may come only twice a month. In hot semiarid areas, what water there is evaporates freely, and during years of prolonged drought only the great rivers continue to flow. In Africa, abundant rainfall in the highlands of the humid Tropics feeds the Nile, and so the Nile's northward flow through the desert never fails. To a lesser extent, this also holds for the eastward flow of the Niger River, its sources sitting in the hills of Guinea and Sierra Leone, near the western coast of Africa. By contrast, the Senegal River practically dried out during the severe Sahelian drought years, in particular in the early 1980s. In the Middle East, both the Tigris and Euphrates arise in the mountains of southeast Turkey, and their strongest flow comes between March and June. That supply of water made it possible for agriculture and civilization to develop more than six thousand years ago in *Mesopotamia*, the region between the two rivers, once Sumeria, Assyria, and Babylon, today Iraq. How much have the rivers' vagaries, catastrophic droughts, and floods shaped the course of history? Some archaeologists argue that a three-centuries-long drought caused the decline of the Akkadian civilization around 2150 B.C.; thick layers of dust are found at the bottom of the Gulf of Oman, in sediments corresponding to that epoch. Other specialists wonder whether salinization of the land was not a factor as important as climate fluctuation—for example, in the vicissitudes of the Harappan civilization of the Indus River valley. The temptation to blame climate change for everything has to be resisted; societies and civilization have their own dynamics, and the course of history depends also on factors such as the discovery and adoption of new food grains, domestication of animals, development of technology, and tactics for plunder and war.

Drought can also strike in the normally humid temperate zone, indeed much more severe drought than made the headlines in Western Europe in the summer of 1976 and in 1989–90. The science known as *dendroclimatology* (i.e., the analysis of tree rings corresponding to the annual growing season) provides long series of data on growing conditions. These are easy to see on the stumps or trunks of trees that have been cut (for example, cross-sections of some very big sequoias, exhibited in various natural history museums of the world), but specialists can use thin "pencils" extracted from the trunk of a living tree without damaging it. Fascinating results emerge regarding the hydrological history of the mid-Atlantic coast of southeastern Virginia and North Carolina, based on analyses of growth rings of bald cypress trees several centuries old, covering the period from 1185 to 1978.[4] The analysis reveals three extremely severe and prolonged drought periods: 1562–1571, 1587–1589, and 1606–1612. This discovery seems to have solved the long-standing mystery of the "lost colony" founded on Roanoke Island (in North Carolina, not to be confused with the Piedmont town of Roanoke, Virginia) by English colonists in 1572 (with no news after August 22, 1587). It seems that these colonists arrived at just the wrong time, as did those who landed on the James River further north, founding Jamestown in 1607. Of the 6,000 persons who made that crossing, only 1,200 were still alive in 1625. Most of the others died from the effects of malnutrition: a terrible death rate, almost as bad as for the slave trade of the next two centuries. Nor should one believe that the Native Americans, presumably better adapted to their environment than the English newcomers, did much better. A Spanish Jesuit missionary, Juan Batista de Segura, visiting the region in 1570, reported that the Croatan (Algonquin) Indians, who practiced farming in addition to fishing and hunting, had been suffering from famine for the previous six years. Critical drought can strike even as green and wet a land as Virginia. Still, the 1562–1612 crisis appears to have been unique over the past eight centuries.

We need water to drink, of course, but that's not all. In Africa and Eurasia, the rapid success and overriding expansion of societies based on agriculture ensues from the systematic exploitation of water in food production. Still, how is it that American civilizations, many of them practicing agriculture well before the arrival of Europeans, did not succeed in resisting the European invaders? How is it that the natives of California, one of the most fruitful regions of the planet, hardly began to grow crops? Many have asked these difficult questions, and Jared Diamond has given some provocative answers in his books *The Rise and Fall of the Third Chimpanzee* and *Guns, Germs, and Steel*, which have very much influenced my thinking in what follows.[5] Why should men have abandoned hunting? Farming is fatiguing, more work than picking apples, blackberries, or wild mushrooms, a lot less fun than hunting or fishing. Writing this, I must admit that today some people disapprove of hunting, and anyhow, in our "advanced" societies, hunters, fishers, herders, and farmers only constitute a small percentage of the population; and many people spend their time staring at a screen, as I am in "writing" this, both for work and for "pleasure." Is that progress?

At the dawn of civilization, agriculture enormously boosts the efficiency of human toil. When you've picked all the berries that the bears have left, you have to travel on to look for others. And prey does not sit and wait for the hunter; those trustful creatures that did quickly disappeared. With their roots in the ground, plants can't run away; and photosynthesis harnesses the free resource of the Sun's rays to transform water and carbon dioxide into organic molecules that then can be "burned" to produce energy. That is the key to all animal life on dry land. Practicing agriculture, humans water and irrigate the plants that they have chosen, taking advantage of free sunlight and carbon dioxide in the air, all this on land they control, from which they try to exclude competing animal species.

Agriculture's expansion goes along with the domestication of animals. Crops convert solar energy into food, a fuel that can be "burned" to provide energy not only for direct use by us humans but also for other animals: donkeys, oxen, buffalo, and horses, who have served humans for thousands of years and still are on the job in many areas of the world. A large part of farm production always went for fodder, i.e., for the fuel powering the engines constituted by pack and draft animals. Still, once a tribe took up agriculture, it could feed a much larger population than could survive on hunting and gathering alone. With their energy resources vastly increased in this way, such emerging societies could accumulate surpluses for stockpiles[6] or for trade, build cities, maintain specialized if not directly productive professions: not only warriors and priests but also architects, surveyors, bookkeepers, astronomers, and poets. And they could invent writing, the beginning of history. With larger and larger populations and with full-time warriors, herding and farming tribes could take over the best land, pushing aside those living in the old hunting-and-gathering way.

The only ones who could effectively resist this expansion and indeed make their own law were the professionals of plunder—raiders on horseback, or on rivers and seas. But plunder only pays in a world with societies that accumulate wealth thanks to the sweat of peasants and artisans. Today, tourists and French natives alike think of Normandy as part of the French countryside. But at the entrance to the archaeological crypt underneath the square in front of the cathedral Notre Dame de Paris, the visitor can view a (restored) plaque in memory of Count Eudes of Paris, Bishop Gozlin, and eight other brave Parisians—Hervi, Arnoud, Seuil, Joubert, Gui, Hardré, Aimerel, and Gossonin—who died in February 886 defending the tower of the Petit-Pont against Norman raiders, on their way to plunder the rich provinces of Burgundy and Champagne further upstream. Viking raiders often used the easily navigable Seine River, and in 885 they came 30,000 strong in 700 boats. But the Vikings were also traders, and for centuries and indeed millennia rivers have been major trade routes, making it possible for some agrarian societies to specialize and to exchange their surpluses against other commodities, whether it be amber collected along the shores of the Baltic Sea, incense from Arabia, or the productions of urban craftsmen.

THE NILE

Growing along the great rivers, the first empires began improving them, putting armies of slave labor to work reinforcing their banks, digging wells and canals, generally seeking to take maximum advantage of the usually dependable though sometimes vacillating resources in the water table and the river flow. Along the Nile, the Egyptians could count on the one hand on regular year-round flow coming from the White Nile, on the other hand on floodwater and fertilizing silt deposits during the summer when the swollen Blue Nile and Atbara Rivers raise the Nile River's level. The White Nile hardly varies at all, its waters supplied by rain of the equatorial zone, its flow effectively regulated by the large volume of Lake Victoria. Conversely, the Blue Nile and the Atbara originate in the highlands of Ethiopia where strong rains fall only during the summer months, and practically none from October to the following April. As a result, flow in the Blue Nile increases by a factor 50 from April to September, and it supplied more than two-thirds of the total Nile flow during the August floods, when water could be diverted for irrigation. Silt and clay transported from Ethiopia by the tumultuous Blue Nile flow was deposited along the banks of the Nile and maintained the fertility of the lands of Egypt surrounded by desert. Approaching the Mediterranean Sea more than 6,000 km (3,600 mi.) north of the point where it leaves Lake Victoria, the river separates into several branches over 280 km (170 mi.) of coastline, from Rosetta,[7] east of Alexandria, to Damietta, west of Port Said. Even here, the Nile still carried a copious load of silt and clay, and over the millennia these sediments built up the extraordinarily fertile and densely populated delta. When exactly did this saga begin? Sea level rose by nearly 130 meters at the end of the ice age, and 5,400 years ago it may have reached a level more than 4 meters above the present level. Could this rise account for the biblical story of the Flood? When ancient Egyptian artifacts are dated using radioactive carbon-14, some of those from Upper Egypt are over 10,000 years old, but nothing

older than 4,600 years old has been found in the delta. French paleoclima-
tologist Jacques Labeyrie argues that the Nile delta as well as the lower
Chaldean plain only emerged from the sea at that time.[8]

All that has been described here in the past tense. Since the completion
of the Aswân High Dam in 1970 and the filling of the reservoir behind it in
the 1960s, the Nile must pass through the 157 cubic kilometers of man-
made Lake Nasser, a volume that corresponds to five years' average river
flow. The dam produces electricity and has made it possible to irrigate new
lands in the desert, without cutting off the supply of water for irrigation
downstream. However, no longer does the river deposit loam and clay
eroded from the Ethiopian highlands, fertilizing the lower Nile banks during its summer floods. Instead, these rich sediments settle behind the dam,
of no use to anyone. In a few decades, they'll have filled up Lake Nasser,
rendering the dam ineffective. At the same time, land below the dam progressively loses its fertility and, with the slower flow of the river, the serious
parasitic disease schistosomiasis becomes more prevalent. In the delta, saltwater intrusion in the water table has increased. Moreover, reduced transport of suspended minerals to the Mediterranean impacts algae and other
forms of marine life, and the catch of fish has suffered as a result. Can production of two million kilowatts of electricity and irrigation of the surroundings of Aswân compensate for all this damage?

WATER'S REVENGE

With the first chapters of the agricultural and hydrological revolutions, humanity commits itself to the treadmill race to keep food production in step
with growing population. Making their start with local and temporary control of their water resources, early agrarian societies grow and expand their
territories, forcing nonpractitioners of farming to flee to poorer land, or
simply massacring them. Just think of the lost cultural diversity! In Europe,
only one non-Indo-European language (Basque) is still spoken today, and
in the Americas, where dozens of distinct language families existed as little
as 200 years ago, few are left as living tongues. As more and more groups
settle down to practice farming, as their populations grow, so grow the
number of people vulnerable to any failure of the water supply. Scarcity
and famine recur all through human history up to quite recent times, in

Europe as well as in Africa and Asia. Of course, in many parts of the world today, dams and river works guarantee fresh water and maintenance of food production even in drought years. But only a little over fifty years ago, the population of India was threatened with penury whenever the monsoon failed, and even in 1967 (not a drought year) rice-free days were declared as a sign of solidarity, although that didn't prevent restaurants catering to tourists and wealthy Indians from serving heaping plates the other days! Has the "green revolution" increase of agricultural productivity put an end to these problems? Many populations remain vulnerable in the poorest countries of the semiarid zone, especially across the African Sahel from Senegal to Somalia. Although modern transport together with a relatively recent surge of human solidarity bring food aid to populations in distress, famine, resulting from floods and drought often aggravated by so-called civil wars, still afflicts millions of people in the world.

The vulnerability of agrarian societies takes another form in the great river basins of Asia, especially in China, where huge numbers of people live in the immediate vicinity of the rivers and depend totally on them for their livelihood. Those rivers often overflow their banks, sometimes with catastrophic results. In northern China, the Huang He (Yellow River) floods of the past raised its banks and enriched the land, in a sense the river creating the entire plain of northern China with a population greater than 200 million. In their centuries-long quest to master the river floods, the Chinese transformed the river banks into still higher dikes, towering in some places as much as 10 meters above the plain. With dikes so high, any breach can have disastrous consequences, the outpouring water with its rich mud washing away herds of cattle and drowning hundreds of thousands of people. Loss of life exceeded a million in 1887—and also in 1938, when the Chinese army deliberately breached the dike, desperately trying to stop the Japanese invasion. More recently, floods took two million lives in 1959. However, although the Yangtze (Chang Jiang) floods of August 1998 drove some 14 million people from their homes and ruined 11 million acres of farmland, loss of life was limited to 2,000—still too much, but much less than earlier.

To prevent recurrence of such catastrophic floods and at the same time harness the power of the great rivers, the Chinese have built many dams and hydroelectric plants, especially in the Yellow River upstream of the loess plateau. In this policy, China might appear to have followed the example of the USSR, basing its industrial development on gigantic dams and hydro-electric power plants. However, in contrast to the Aswân High Dam behind which the Nile River's silt accumulates, the dams of the upper Yellow River retain only relatively clear water, regulating the river's flow before it reaches the most densely populated areas. Also, it should be remembered that China has an ancient tradition of river works and canal construction. As already noted, the Huang He changed course many times, notably in 1851–52, and even more radically in 1324. But the first Grand Canal, built by millions of men and women, was completed in 605 A.D.; the second, following the Mongol conquest, in 1289. The principal objective appears to have been to ensure secure and convenient transport of cereals from the well-watered regions of the south to the semiarid north, where the Mongols and their successors had set up the capital Peking (now Beijing), with its myriad civil servants, scholars, craftsmen, and artists.

More recently, the government of China has undertaken the gigantic Three Gorges dam project on the Yangtze Kiang, more usually called (as earlier noted) Chang Jiang or "Long River" by the Chinese. Indeed, the Yangtze (6,000 km or 3600 mi. long), third greatest river in the world, has an average flow of 30,000 cubic meters per second, a trillion cubic meters per year. Its vigorous flow is generally more regular than that of the Yellow River, but the strongest Yangtze floods have nonetheless taken hundreds of thousands of lives, especially along those stretches where the river flood plain narrows and in passages through gorges where the swiftly flowing water can only rise dramatically. Riverboats have long plied the Yangtze, and diversion of its water for irrigation began ages ago; and construction of dams on its tributaries to control the river's flow through the densely populated plain began years ago too. With the Three Gorges Dam, electric power production should ultimately reach 12,000 megawatts for the Szechwan region. With the construction of enormous locks and ship elevators, 10,000-ton cargo ships should be able to reach at least as far upstream as Chongqing, the city that was the capital of nationalist Chinese resistance against Japan in World War II. However, the project involves flooding hundreds of towns and villages over a large area, forcing the displacement of more than a million people. For this reason, and also because of ecological impacts as well as the risk of earthquakes, many in China and elsewhere dispute the Chinese government's wisdom in pushing through this project. Can the long river really be mastered?

Population growth entails both increasing appropriation of fresh water and more and more discharge of "used" water—human and animal sewage, microbes, etc. Along rivers, downstream water users get upstream users' waste; and if the river overflows its banks, even well water may be polluted. The worst floods of China took their toll not only by drowning and famine but also by epidemics. Throughout the humid Tropics of Asia, from the Indian subcontinent to southern China and Indonesia, water's revenge takes the form of parasitic diseases such as schistosomiasis, bacterial diseases such as cholera and typhoid, and (by way of water-loving mosquitoes) such scourges as malaria, yellow fever, and dengue, even in the absence of floods. Intensive agriculture takes maximum advantage of the abundant water resources, but by the same token it creates and supports high population densities in which diseases easily spread.

Population certainly did not reach many hundreds of thousands in the earliest cities of Mesopotamia, but hygiene was taken seriously in those centers of power and wealth. Further east, Mohenjo-Daro, metropolis of the Indus valley civilization that flourished 4,000 years ago, boasted both public and private wells providing clean water, and a system of vaulted culverts for removal of sewage. The Romans built aqueducts and sewers for the cities of their empire, the first closed sewer, the *cloaca maxima* of Rome, dating back to 500 B.C. Even though Paris, the Romans' *Lutetium*, was only a minor provincial town with population less than 10,000, the Romans built a 12-mile-long aqueduct (from Arcueil, a southern suburb) as well as the baths whose ruins can still be seen in the Latin Quarter. After the Romans left, things went down the drain (or rather, they no longer did): in the Middle Ages, filth flowed down the middle of the streets of Paris!

Contaminated water causes cholera, a disease of diarrhea and vomiting that must have afflicted the inhabitants of the waterlogged lands of the Ganges delta, now in Bangladesh, for centuries and millennia. With the vast increase of trade in the nineteenth century, ships with their sailors and their drinking and bilge water carried the disease across the seas, and it spread easily in the filth and squalor of the teeming cities. The cholera pandemic that started in India in 1826 killed many thousands in different European cities, reaching New York (by way of Canada) in 1832. Thousands of New Yorkers died, one in fifty of the total population, and as many as 100,000 fled to Greenwich Village, then a countryside refuge outside the city.[9] In 1857 another cholera pandemic struck southern

France, in particular the Mediterranean port city of Marseilles, and it reached New York in 1865. Earlier, in 1854, English physician John Snow had observed that all those who came down with cholera had drunk water from the same pump on Broad Street in the City of London. Although he had no idea of what in the water could be causing cholera, Snow put an end to the epidemic simply by removing the pump handle, and in so doing he identified the mode of transmission of the disease. It was later, in 1883, that German bacteriologist Robert Koch,[10] on mission to India and Egypt, identified the guilty microbe *vibrio cholerae*, although Filippo Pacini (1812–1885) already saw it in Florence in 1854. A new and dangerous strain, *El Tor*, appeared in 1937. Rapid rehydration (replacement of fluids and salts) of cholera victims is absolutely essential and effective, and the disease can usually also be treated with antibiotics. However, sanitation and water purification constitute the first lines of defense against cholera. Vaccination can be of value of travelers to risk areas, but it is ineffective against some strains, and when effective, it is only for several months at best.

Although Paris chemist Claude Louis Berthollet discovered sodium hypochlorite and commercialized it as a laundry bleach starting in 1790, nearly a century passed before systematic use of chlorine for disinfection of drinking water. Chlorination has contributed enormously to public health in the world, but cholera still takes too large a toll in the poorest countries. Worse, in 1990–91 a cholera epidemic took more than 16,000 lives in Colombia and Peru, in part because chlorination of the public water supply was stopped![11] Despite its contributions, chlorine today is under attack in the United States and other rich countries of the world. There is no doubt that some chlorine compounds harm the environment; disinfection is, after all, an aggression against certain life forms, an offensive defense against disease-causing microbes. Furthermore, chlorine reacting with organic material present in water can form compounds of the organochlorine family, in particular trihalomethanes (THMs), feared as potentially carcinogenic. At what concentration does the risk become significant? If one sets too low a limit on the admissible THM level, the microbial risk may rise too high. Is it true, as some argue, that with modern technology, the effectiveness of disinfection by chlorination can be maintained without producing too much THM? But what is too much? Every mention of the word *cancer* naturally frightens people, but the THM threat must be compared with the very real risk of bacterial disease. An undiscriminating antichlorine crusade could be catastrophic for public health.[12]

Drinking Water for the Cities

Once you have air to breathe, once you have enough warmth (but not too much!) to keep your body temperature close to 37°C (98.6°F), the next most urgent vital need is water to drink. But water can carry disease; the problem is to find safe drinking water. And this problem becomes increasingly difficult to solve as the growing populations of the cities require more water than can be supplied by wells and springs, and as the growth and industrial activities of these same populations lead to increasing pollution and contamination of rivers and lakes, and even of the subsurface water table. The problem arose long ago, and it grew along with the cities. Today's average Parisian "consumes" some 250 liters of water every day, water for drinking (only about 3 liters), for showering or bathing, for washing salad greens, dishes, clothing, the sidewalk, etc., for various trades and industries, and for fighting fires. At the time of the French Revolution, water supply was less than a liter per day, apart from soiled Seine River water.

Although Seine water was quite pure in 358 A.D., the Romans brought spring water to Lutetium (Paris) from the hills to the south (fig. 11.1). Following the Arcueil aqueduct's destruction by eleventh-century Norman raiders, Parisians again had to rely on water from wells or from the Seine until Philippe Auguste (king of France from 1180 to 1223) had new aqueducts built to supply three public fountains. One of these, the Fontaine des Innocents, next to the very central Châtelet-les Halles express metro station, continues to attract Parisians, suburban commuters, and tourists (although not to drink its water). By 1669 the city had thirty-five fountains fed by waters coming from the surrounding villages of Rungis, Belleville, and Pré-St. Gervais, now within the Paris city limits. As the city grew, its waterworks had to improve, and the nineteenth century saw the building of the Canal de l'Ourcq, installation of increasingly powerful pumps along the Seine, and the drilling of artesian wells in the neighborhoods of Grenelle, Passy, and the Butte aux Cailles.[13] With new sources of water (la Vanne) and the new Arcueil aqueducts, tap water could finally reach higher than the first floor; it became fashionable to live upstairs, especially after elevators arrived. Water meters made their appearance, and the city authorities began to set standards for water quality: six parameters in 1885, twenty-five in 1964. Public health authorities must show that the norms are indeed satisfied; measurements are posted, for example, at the suburban express metro station where I take my train to get home from my lab at the Ecole Polytechnique.[14] The European Commission (executive arm of the European Union) now has a list of over a hundred parameters in its norms, including not only turbidity (cloudiness) and microbiological parameters but also acidity (pH), radioactivity, nitrate contents, different metals, minerals, dissolved gases, pesticides, etc.

French consumers insist on the quality and taste of their drinking water. Demand rose to 226 liters per person per day in 1975, and 280 in 1982, but seems to have stabilized since then. In many areas of France,

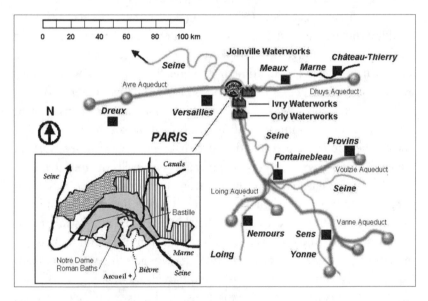

FIGURE 11.1 The Paris water supply. The circles show areas where spring water is tapped for the aqueducts leading to Paris. In addition, the waterworks outside the city limits at Ivry, Joinville, and Orly treat river water from the Seine and Marne. The inset map (approximately 21 km or 13 miles across) shows water distribution districts within the city: Avre spring water in the northwest, Dhuys spring water in the northeast supplemented by treated river water from Joinville or Ivry, Vanne and Loing spring water (light gray areas), supplemented by treated river water from Orly (unshaded areas within Paris). The Romans tapped spring water near present-day Arcueil and conveyed it to the baths of Lutetium.

ground and spring water still satisfy 80 percent of the need. This water from underground is nonetheless "meteoric" water, depending on fairly recent rains but regulated and filtered by layers of soil. From the legal point of view, France depends on the 1992 Water Law, according to which each of the 30,000 points at which underground water is tapped should have a surrounding perimeter of protection; but many of them still are not so protected and, moreover, in many (most?) cases pumping is not metered and no one has a clear idea of how much water is being withdrawn from the water table. Even so, spring water does not cover the total need. In Paris and in other cities such as Lyons, Toulouse, Poitiers, Nantes, and Rennes, less than 45 percent of the water supply comes from underground springs, most being diverted from rivers or from reservoirs built to regulate river flow. After the record Seine flood of 1910,[15] dams were built to retain as much as 800 million cubic meters of water, about what the Paris metropolitan area consumes in a year. Use of river water, from the Seine and the Marne in the case of Paris, has by the nature of things the drawback that it includes both "used" (possibly polluted) water discharged by upstream users and poorly controlled runoff water—in particular, runoff from agricultural land containing nitrates (from chemical fertilizers), animal wastes, bacteria, pesticides, phosphates, and so on. River water must be treated more than spring water, but safety of spring water cannot be assumed without checking, and treatment may be needed. Modern water treatment involves several stages: decanting, filtering (in particular by active carbon, filtration using ultra fine membranes), aeration, and, finally, disinfection, for which chlorine or ozone usually play an essential role.

New York, like Paris, gets its water more than 100 km (60 mi.) upstream, indeed even from twice as far away (figs. 11.1 and 11.2). Los Angeles imports water from farther away still, from the mountains, and partly from outside the state of California. Aqueducts came to play a role New York's water supply much later than in Paris, but that role quickly became preponderant. Local Hudson River water, salty because of tidal inflow, never was an option. Early inhabitants—first Native Americans, later Dutch and English colonists—used water from springs and streams. Private wells were dug as early as 1658, and the first public well in 1677. Wells still supply water in southeast Queens (far outside New York City's seventeenth-century limits), but today such groundwater amounts to only 2 percent of the city's water supply. In 1755, Bethlehem, Pennsylvania, put into service the first pumped water supply in the thirteen colonies, and a gravity-fed piped system operated in Providence, Rhode Island, starting in 1772. Christopher William Colles built a waterworks engine in New York City in 1774–75, but British occupation troops destroyed it in 1777.[16] New York City's population grew from 60,000 in 1800 to one million in 1878, much of the economic growth following the opening in 1825 of the Erie Canal, which completed a navigable waterway from New York Bay up the Hudson River and thence west to the Great Lakes.

Awareness of the need for clean water grew even faster following the patenting of the flush toilet in 1819, the 1832 cholera epidemic, and the great fire that destroyed much of the city in 1835. Work on the Croton River Dam and Aqueduct began in 1837, and by 1880 New Yorkers were consuming 100 gallons per person per day, the highest water-use rate in the world. Later New York City had to go to the Catskills, further upstate, flooding many country villages, and when that no longer sufficed the city had to negotiate water allocations from the Delaware River, an interstate waterway, with the state of New Jersey, finally obtaining them only by a decision of the U.S. Supreme Court.[17] The United States and four states adopted the Delaware River Basin Interstate Compact in 1961, and the system was completed in 1964, just in time for the 1964–65 Northeast drought. New Yorkers' water use peaked at about 212 gallons (800 liters) per person per day in 1990 but has decreased since then, thanks in part to expansion of metering and installation of water-saving plumbing devices. That's still over twice the corresponding average for France!

CLEAN BUT TOXIC WATER

In the developed countries, nearly all city-dwellers have drinking water on tap from a public distribution system, and so do many rural inhabitants. Almost everyone has access to reasonably safe clean water.[18] In the poor countries, clean water cannot be taken for granted. Although a majority of city-dwellers (at least those who live in buildings rather than shanty-towns) have access to safe water supplies, less than half do in rural areas. Moreover, considering the accelerating population growth of nearly every Third World metropolis, and the lack of resources to develop already poor

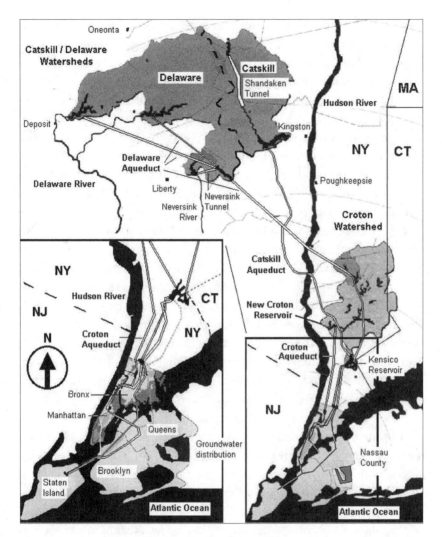

FIGURE 11.2 The New York City water supply. The northern limit of the Delaware watershed is 125 miles (about 200 km) from City Hall in lower Manhattan. Croton water goes primarily to the shaded areas of Manhattan and the Bronx, often also to lower Manhattan. Most of the city uses water from the Catskill and Delaware watersheds, although the darkly shaded part of Queens uses groundwater. Distribution involves an extended and complex system of reservoirs, above-surface aqueducts, and tunnels. (Adapted from maps of the New York City Department of Environmental Protection)

public services, more and more people have to use and drink dirty or doubtful water. As a result, water-related diseases, gastrointestinal infections, diarrhea, etc. are on the rise. The situation in the countryside is even worse. In Burkina Faso (in the Sahel region of Africa), 75 percent of the traditional wells are contaminated, and *all* the ponds. That may be an extreme case, but every day some 600 million people in the world suffer from gastrointestinal infections, most of them in the poor countries, and the affliction causes some 20 million deaths every year.

The cause of public health in the world requires constant effort to find clean water and to keep it clean, but this is not enough. Disease caused by contaminated water once took a terrible toll on rural populations in the lower delta of the Ganges and Brahmaputra Rivers. Starting in the 1970s, Western humanitarian organizations, working together with the public health authorities of Bangladesh and of India's West Bengal state, initiated an intensive effort to find clean water, to bring it to the villagers, and to convince them to abandon their tradition of using surface ponds and shallow wells. They succeeded and promoted drilling into an aquifer with abundant clean water between 7 and 100 meters underground. More than two million cheap "tubewells" (steel tubes about an inch in diameter) were put into service. And since then, partly because of this germ-free water, partly because of other health measures, the incidence of water-related bacteriological disease has dropped significantly. Tragically, however, the solution brought with it another set of tribulations. Beginning in the 1990s, new health problems increasingly afflicted the people of this region, with a dramatic rise in skin ailments and in melanomas and cancers of internal organs. These result from chronic exposure to arsenic, for which the most serious health effects only appear after ten or fifteen years. Systematic tests of the tubewell water showed, twenty years too late, that in most locations the water's arsenic content is as much as 20 or even 200 times higher than the limit of 10 micrograms per liter advised by the World Health Organization.[19] This arsenic "pollution" of the water is a perfectly natural consequence of the geology of the region. The problem has been observed elsewhere in the world, notably in Taiwan, Chile, and Argentina, and in Utah in the United States, but never on this scale, affecting tens of millions of people. What next? The challenge is to find a way of eliminating the arsenic, easily, at low cost, preferably using local materials, and to retrofit quickly the millions of tubewells in use. Scientists and engineers in many countries are working on it, and international organizations such as the World Bank are providing some funding.

WATER DOWN THE DRAIN

Even (or especially) in the rich United States, urban and suburban inhabitants often complain about water shortages, but what do they do with their water? Practically all of the water "consumed" by city-dwellers goes down the drain, returned still liquid to the environment, but with its quality degraded. By the nineteenth century, the once-clear little Bièvre (beaver) stream of Paris had become an ugly stinking cesspool, flowing with wastes discharged by dyers and other polluting craftsmen; it finally had to be covered up from the south of Paris to the point where it enters the Seine. After allowing sewage to flow in the streets for centuries, the authorities finally built a citywide sewer system: 26 km of sewers in 1805, 155 in 1854, 1,650 at the start of the twentieth century. This system got the sewage out of Paris, but it still returned the soiled water to the Seine River, to flow past Rouen and other towns. One town's downstream is another town's upstream. How can upstream dwellers keep a clear conscience when they send waste water, in which they would never swim, into the river or out toward seaside beaches? Downstream inhabitants can and do complain. The only downstream for New York City is the ocean, but New Yorkers like to go to the beach. Also, fortunately, increasing numbers of citizens can and do speak for the protection of aquatic ecosystems, the whole delicate network of species living in or near the water. Even when water pollution is not, strictly speaking, toxic, the consequences can be dramatic. Phosphates lead to *eutrophication* of some lakes and streams; their fertilizing action promotes excessive growth of certain life forms (microbes and plants), exhausting the dissolved oxygen in the water to the point where few fish survive.

Modern urban water supply involves fairly expensive collection and treatment of the water, and elaborate distribution systems. Why bother to supply drinking-quality water at such cost, when most of it ends up being used to flush the toilet, or to wash the kitchen floor or the street? But in what sense is the cost high? In one year, the 1990–91 cholera epidemic cost Peru the equivalent of more than ten years' investment in water treatment and distribution. In the former Soviet Union, as the state withers away under conditions that were not imagined in Lenin's time, water-borne diseases have reappeared. If you visit Russia, it's advisable to get vaccinated against diphtheria, and if you decide to take a swim in the Moscow (Moskva) River (as many Muscovites do in the hot summer), try not to swallow the water.

Of course, in riverside cities untreated water can be used to wash the streets and to fight fires, and in New York, fire hydrants are also put to use on hot summer days to provide inner-city showers for kids who can't get to the beach. It's been proposed that such a dual water-supply system should be extended into apartment buildings and private houses, providing separately on the one hand expensive drinking water, on the other less elaborately treated but still safe water for flushing toilets and for cleaning. Questions of practicality and risk can be raised, and there are alternatives. In many places, waste water is recycled for industrial applications, for irrigation, or for watering parks and gardens. In Japan, water already used in the kitchen or for soaking in the bath is often reused to flush the toilet. More generally, there has been an effort, at least in the developed countries, to treat waste water, often in large-scale installations, before returning it to rivers, lakes, or the sea. In 1989, such treatment was applied to 98 percent of the waste water in Denmark, 52 percent in France, 73 percent in the United States, and 39 percent in Japan. The ultimate objective in rich countries is to reach a situation where no one need fear swimming in the river, even just downstream of the release of the treated waste water.

Even with such water-treatment facilities, the paved impermeable surfaces of cities create special problems. With very little if any infiltration possible, violent downpours go almost entirely into storm runoff carrying all sorts of trash, dust, hydrocarbons, and other dry and greasy deposits, from the sidewalks and streets and even rooftops into the sewers and rivers. The resulting short-term pollution can lead to massive fish-kills in the river. Provided that the effects remain short-term and that the aquatic ecosystem can recover, so much the better for city-dwellers with that much less dust in the air they have to breathe. Indeed, systematic spraying of the streets every evening can play the same role and reduce the risks associated with storm runoff. It was said that this was an important part of keeping radioactive dust out of the lungs of the inhabitants of Kiev (capital of Ukraine) in the years following the 1986 Chernobyl nuclear accident, but many cities carry out street-spraying in hot summer months. In any event, such practices implicitly assume that the different wastes do less harm when gradually released to the water than in the air and in storm runoff.

WATER FOR FOOD

Clean drinking water must be supplied to the cities; city-dwellers and sub-urbanites should learn not to waste it, leaks and other losses in the distribution system must be reduced, and water-treatment plants must make the waste water clean again.[20] That's all very well, but in the world as a whole, 69 percent of the fresh water used one way or another by human beings goes to agriculture, to grow cereals, vegetables, cotton, and animal fodder; in the Near East, 91 percent, in Central Asia, 95 percent. In North America and Western Europe, it is true that less water is used for irrigation than for industry and the generation of electricity. What sort of use? In France, the cooling towers of electric power stations require more than 26 billion cubic meters of water per year. That's six times more than used for drinking water! Whether they use fossil fuels or nuclear reactors, these power stations warm the cooling water, some of which evaporates and most of which returns warmer to rivers or the sea. For the most part, that "thermal pollution" is wasted; in a few cases the warm water goes to greenhouses and other uses. As for agriculture, France usually can count on rain, and irrigation consumes only 5 billion cubic meters per year, 12 percent of the total water withdrawal. For Western Europe as a whole, irrigation consumes 36 percent of a total freshwater withdrawal of 250 billion cubic meters annually. Still, even in this highly industrialized part of the world, agriculture really consumes much more when one adds in the evapotranspiration of water from farmland. Most of that water, withdrawn from the water that falls on the land, goes neither into infiltration nor into runoff; it returns to the atmosphere to be blown away in the wind. While paved cities return too much, agricultural lands return only 25 percent of their water inputs to runoff. Outside of Western Europe, irrigated agriculture dominates. In Asia, from Korea to India, irrigation accounts for more than 90 percent of the water withdrawals (1,210 billion cubic meters per year). For the world as a whole, agriculture is the biggest water user by far, and saving water depends in the first place on more efficient use of water in agriculture.

Apart from those who lose their way in the desert, hardly anyone dies of thirst as such. Death strikes early in poor countries, but more by starvation or by disease, resistance to disease often being reduced by malnutrition. In some regions of the world, farmers and herdsmen manage very well with the water from springs, from rain, and from runoff in their local watershed. In many regions, however, they depend on a water supply from outside. In dry areas, without a river like the Nile or the Euphrates nearby, and without long canals provided by a strong state, they must learn how to catch and store rain when it falls. In Israel's Negev Desert, visitors can see small dams that the Nabateans built 3,000 years ago to trap rare runoff in the *wadi* before it disappeared in the rocks and sand, and covered cisterns that kept evaporation losses low. In other arid or semiarid regions, along the banks of great rivers, ambitious waterworks, dams, reservoirs, and irrigation canals have been created to increase agricultural productivity, usually under government authority and funding. Major projects of this sort radically changed the conditions of farming in the Missouri River basin in the United States, in China, and along the principal rivers of Russia and the Ukraine. They have not always been successful, however. Indeed, sometimes the net effect has been negative—for example, in the excessive dependence on irrigation in the arid "new lands" of formerly Soviet Central Asia, drying up the Aral Sea. In more humid regions, waterworks, ranging in size from local private constructions to much larger public systems, play a vital role in controlling stream flow, supplying water to cities and also to some extent irrigating cropland. All in all, agriculture is the principal water user in the world, especially if the entire available freshwater resource is considered and not just the water distributed through private or public distribution systems.

Most of the water used by agriculture returns to the atmosphere by evapotranspiration, the exact amount depending on the plant, the state of the soil, and weather conditions (fig. 6.1). Some crops use more than others: corn pumps back to the atmosphere enormous amounts of water from the soil, reducing its moisture. In times of drought, some argue that it would be better to switch to other less demanding crops rather than to ask for water for irrigation; generally, however, irrigated agriculture uses water more efficiently, and corn does not come out so badly in terms of nutritional value per amount of water consumed. Apart from evapotranspiration of water by crops, water is lost by evaporation from leaf surfaces and from the soil if they stay wet because of excessive rain, watering, or irrigation. In those parts of the world where every drop counts, for example in Israel, drop-by-drop irrigation techniques keep losses to a minimum, maximizing profit from the dearly won water.[21] What prevents more people from adopting such systems? Waste of water is enormous in France and in much of the United States. Is the price too low? According to one French water resources manager, "Only high-priced commodities are respected." Why should farmers change their habits and switch to less thirsty crops, when extra irrigation water costs only about 2.50 Euros (or dollars) a cubic meter? Should the price be doubled? As for city-dwellers, even with relatively cheap water, drop-by-drop watering systems can reduce the work of carrying water needed to keep alive flowers on Paris balconies.

Not only is agriculture the biggest consumer of water, it also is one of the biggest sources of water pollution, discharging huge amounts of waste: excessive applications of fertilizer and pesticides, piles of pig and cow manure (in the United States, 1.6 billion tons of excrement per year, half of it from feedlots). With hardly any attempts to manage them, such wastes end up in the rivers or in the soil and the water table. In Roscoff, in an area of Brittany in western France where the mild climate makes possible three crops a year, no one drinks the tap water because of its high nitrate content due to massive fertilizer use. In France, agriculture has an enormous responsibility in the degradation of water quality. The pollution tax rate is far too low to persuade producers of pork to install the equipment needed to collect and treat the pig manure, and farmers do not contribute substantially to the cost of water treatment. Raw pollution due to dairy and meat producers in France is equivalent to that of more than 250 million people. Should the polluters pay? Who are the polluters? In the rich countries of the world, hundreds of millions of consumers drink milk, eat cheese, beef, chicken, and ham, and they too have to pay their share. We are all, directly or indirectly, polluters, and to cover the cleanup costs, some profits must fall and some prices must rise.

LAND FOR FOOD?

Contradictions abound in the complex relationships between humans, land, and water. Because rivers wreak destruction when they breach their banks, we try to tame them, building dams and creating reservoirs to regulate their flow.[22] But some of those reservoirs permanently flood vast land areas, expelling large populations from their homes. In the southern French Alps, some people still remember the old village of Savines, underwater in the Lake of Serre-Ponçon created by the completion in 1960 of the dam on the Durance River. When the water is clear, you can still see the church steeple. And astronomers who knew the Haute-Provence Observatory before then will tell you that the skies were clearer when no water was available for sprinkling. In some areas of the world, people still accept (or are given no choice but to accept) the exchange of flooding some of the land against more easily manageable water for the rest. Elsewhere, people have taken the opposite course. Waterlogged earth, whether with fresh or salt water, marshes, swamps, and other wetlands have long been considered, by those who would till the soil, as so much wasted land, with the reputation of being a source of disease to boot. Europeans have long drained and farmed many of these wetlands, the Pontine marshes in Italy, those near Dol in France or in the Valais plain in Switzerland, as well as some areas of eastern Europe. In America, the surveying of the Dismal Swamp in North Carolina in view of its (partial) draining counts as one of the major accomplishments of the young George Washington. Much of Holland results from a long effort to reclaim land from the sea, with the creation of polders and the erection of the dam closing off the Zuider Zee, turning it into the Ijsselmeer. In Israel, the marshy Huleh Lake near the Syrian border was turned into farmland. In Paris, only the name of the Marais (marsh) neighborhood survives. And in many areas still, land-hungry expanding cities make plans to drain what marshland remains.

Where once wetlands were the first targets of "reclamation," today, at least in the United States and some European countries, people have begun to appreciate their value in protecting a large variety of living species, in rendering a host of environmental "services." Bacteria living in wetlands contribute to natural elimination of certain pollutants, and wastewater-treatment enterprises now invest large sums in trying to reproduce such processes in "bio-reactors." Oysters, clams, mussels, and other shellfish are fine eating, provided that they come from uncontaminated wetlands. Migrating birds stop over there. And so, today, rather than draining them, planners protect wetlands so that urbanites and suburbanites can have a chance to see this other facet of nature. Some wetlands still subsist in the immediate vicinity of New York City despite the encroachments of airports, oil refineries, seaports, shopping malls, and stadiums. In France, several regional parks preserve what's left of wetlands—in Camargue (the Rhone delta of southern France) and other areas. In Sudan, however, marsh reclamation is still a priority, justified by the need to reduce evaporative losses of scarce water, even though far more water—billions of cubic meters per year—evaporates from the surface of nearby Lake Nasser behind the Aswân High Dam in Egypt.

Extension of agriculture raises the risk of soil erosion. In the temperate zone, erosion represents a serious threat on sloping land, especially in less humid areas where the soil tends to dry out in between strong rains. In southern Europe and elsewhere, wine-growers learned long ago to limit such erosion by terracing the hillsides, an example of the "wooing of the Earth" described by René Dubos,[23] the construction of beautiful humanized landscapes as well as the production of exquisite wines. In the tropical zone, the terraced rice paddies that have replaced other landscapes have their own beauty. In some areas of the humid Tropics, if the protective tropical forest cover is removed, abundant rainfall and excessive runoff can wash away the soil or leach it, transforming it into brick-hard but barren *laterite*. The technique of agro-forestry, farming under a tree canopy, can help to avoid this. Although the shade may reduce productivity of some crops, for others it improves quality. The trees protect the soil during violent downpours and they also provide shelter from the wind, reducing evaporative moisture losses during the dry season. In southern China, the practice of growing tea under trees has saved some forests that otherwise would have been converted to farmland because of their lack of profitability. And in Indonesia, agronomists have succeeded in grafting a *Ceara* rubber tree onto the manioc plant, and the combination protects the soil from strong rain on the one hand, the manioc from wild pigs on the other, while at the same time multiplying overall productivity by a factor 10.[24]

Mosquitoes and Disease

Contradictions abound in the relationship between humankind and water. The scourge of malaria remains the disease taking the biggest toll among humans on this planet. The name *malaria* comes from its association with the presumably bad air of marshland. In fact, as shown a century ago by British physician John Ross stationed in Calcutta, malaria is transmitted by the bite of *Anopheles* mosquitoes. The complete life cycle of the disease-causing parasite was finally understood in the mid-1950s. But long ago, before anyone knew how malaria infection occurred, the elimination of the supposed source of infection was a major motivation for the draining of marshland. Once the role of mosquitoes was understood, once effective insecticides such as DDT became available (and before mosquitoes developed resistance), elimination of mosquitoes became the tactic of choice in addition to the draining of wetlands.[25] After all, malaria still existed in Italy (in particular in Sicily) at the end of World War II; and in the early 1960s, when a plane from Paris arrived in New York, an officer would spray the cabin with DDT before anyone (or any mosquito) could exit. However, the last thirty years have seen growing awareness of the persistent and perverse effects of DDT in the environment, and its use is prohibited in many countries. Now, when American air passengers find themselves being sprayed with insecticide on arrival at certain tropical airports, they sue the airline. And developers in many countries are warned off the wetlands.

Because of the complexity of nature, contradictions between environmental protection and public health cannot always be avoided. Human cultural evolution and medical technology have made enormous progress in the fight against disease, but many of the disease-causing micro-organisms have managed quite well to keep up, developing resistance by way of biological evolution and natural selection. From an environmental point of view, animal wastes should be used as natural fertilizer. In China, such "ecological" methods have taken the form of breeding pigs, ducks, and fish in close association, and also using human feces abundantly. But such ecological promiscuity probably also promotes the formation of new and sometimes potent strains of the influenza virus.

WATER FOR INDUSTRY

In modern countries, industry and power production use large amounts of water, about half of the total withdrawals. However, nearly all the water withdrawn from river flow (for example) for cooling purposes in power plants is returned to the river near the point of withdrawal, with only the temperature changed if all goes well. As for industry, it too returns a large share of what it withdraws, much more than agriculture, but too many industries still return a poisonous polluted flow, what with detergents, hydrocarbons, heavy metals, and toxic organic matter. The classic case, subject of many rock tunes, is the Cuyahoga River, flowing through Cleveland, Ohio, that was plagued by fires starting in 1936 and as late as June 21, 1969. It has since been cleaned up.

And while pollution may not always strike the eye or insult the nose (don't try to taste the stuff!), the insidious accumulation and concentration of certain substances in the food chain can lead to serious illness.

Industrial water pollution is rightly viewed as scandalous because, if a factory's operation is likely to pollute the water it uses, what better place to clean up the water than inside that same factory? Add to the production or assembly line a water-treatment chain to restore the quality of the water before discharging it into the rivers. That problem can certainly be more easily solved than the problem of cleaning up the runoff from the less concentrated activities of agriculture. Things are looking up, it must be admitted. More and more, modern factories recycle the water that they use, and clean up the water they discharge. Would factory owners and operators take the trouble if they were not exposed to public outcry resulting in regulations that they must follow? And what if their products' prices rise as a result? Everything has a cost, and why should all of the sanitary, ecological, and aesthetic costs resulting from a particular industry's pollution, related to the satisfaction of particular consumers' needs or wants, be passed on to the community as a whole? The problem is to price products fairly, taking into account the safeguard of water quality and, more generally, the protection of the environment.

WATER BUSINESS

Most of us count on being able to turn on the tap to get a worry-free drink of water, or to water the flowers precisely when the weather is dry. That possibility depends on a long chain of efforts and technical accomplishments that have a cost and must be paid for. Can there be a "fair price" for water, an absolute necessity of life? But why should water be treated differently from "our daily bread," also a vital necessity? For water as for bread, the gift of God or of heaven must be earned by the sweat of one's brow. In most societies, a deeply ingrained ethos of solidarity requires that we not allow our fellow human beings to die of thirst or starvation, and many governments today provide subsidies to keep down the cost of water or of bread, sometimes counterproductively. Selling water has been a business from way back. In Paris, the last street vendors of (unbottled) water disappeared a century ago; but in many countries they still ply their trade.

The sale of water in its solid state is still going strong, and along most roads in the United States you can buy large bags of ice cubes for keeping cool your picnic chests. That business isn't new either. Telling the story of his "life in the woods" at Walden Pond (only two miles from the highly civilized society of Concord, Massachusetts, west of Boston), Henry David Thoreau[26] described a singular winter harvest in 1846–47: "A hundred Irishmen, with a few Yankee overseers, came from Cambridge every day to get out the ice.... The heap, ... estimated to contain ten thousand tons, was finally covered with hay and boards" to protect it. The organizer of the operation hoped to double his investment, in spite of the 75 percent loss between the lake and the buyers. However, according to Thoreau, the speculation failed; by September 1848 the pond recovered the greater part of the water.

Before the general climate warming that began around 1850, before the intensification of the urban heat island of Paris, the ponds of the Bièvre froze every winter. Enterprising businessmen harvested the ice and kept it cool for sale in the summer. The Bièvre has been covered for many years now, but the Rue de la Glacière ("Icehouse Street"), not far from where I live, reminds us of those times. A few of us still remember the times before everyone had an electric refrigerator, when the iceman came down the street every day to sell blocks of ice for the family icebox, in New York City as well as in Paris. Every so often you read about a project to tow icebergs from Antarctica to the Persian Gulf, where oil is plentiful but fresh water is scarce: but what fraction of the ice would reach its destination?

As already mentioned, bottling spring water is big business. A new potential business would involve selling large volumes of lake or river water. In 1998 the Ministry of the Environment of the province of Ontario gave the go-ahead to a project of the Nova Group in Sault Sainte Marie, to pump some 10 million liters of water per day from Lake Superior, filling up tankers to sell the water in the Persian Gulf. However, the Canadian federal government blocked the project, using as grounds the 1909 Canada-U.S. treaty and fearing to create a precedent for other such sales. In the Near East, water-thirsty Israel has considered buying water from water-rich Turkey. Transport could be by tanker, or perhaps at lower cost by gigantic tugged water bags; an undersea pipeline would be an expensive long-term investment, implying perhaps justifiable pessimism regarding the likelihood of Israel eventually buying water piped in across its land neighbors' territories. In southern Europe, the economically very active Catalonia region of Barcelona needs water, but Spain is already short. The possibility of getting water by aqueduct from the Rhone in southern France has also been discussed, inasmuch as relatively little is being withdrawn from that river's flow. In all these cases, and others, fresh water is a commodity bought and sold—stories to be continued . . .

NUCLEAR POWER AND WATER

Perhaps no pollutant provokes as much anxiety as radioactivity. Radioactive nuclei released when nuclear and thermonuclear bombs were exploded in the atmosphere are still with us, having fallen out from the atmosphere onto and into the water and the soil, being taken up by grass, vegetables, mushrooms, and milk. The many test explosions carried out before the atmospheric test ban treaty of 1963, principally by the United States and the USSR, exposed hundreds of millions of persons to radioactivity, including millions of babies and young children who drank milk containing radioactive iodine-131. For the most part, except for those who happened to be immediately downwind of the explosions, the dose was weak, but few if any knew what they were being exposed to. Bomb tests continued for many years after 1963, but underground.

Bombs should not be confused with the peaceful production of electric power in nuclear reactors, but strong controversy continues regarding nuclear power and in particular nuclear waste. It should first of all be remembered that in "normal" operation, a nuclear reactor should release hardly any radioactivity at all in the air or in the water, indeed in many cases *less* radioactivity than released by a coal-burning power plant, inasmuch as coal often naturally contains traces of radioactive elements. In Brittany (western France), the natural radioactivity level due to the native granite certainly far exceeds the radioactivity released by the nuclear power plants. The problem is not in normal operation, but in what can happen during an incident or accident, or in what people imagine can happen. In France, the national electric power company EDF has many years of experience operating nuclear reactors which produce three-quarters of the electricity in the country, and there never has been anything more than a minor incident. In the United States, the radioactivity released during the Three Mile Island reactor accident in Pennsylvania remained confined within the protective shield and very little reached the environment. The most serious effects—no doubt viewed by antinuclear activists as the most beneficial effects—were psychological, and practically no new nuclear power project has been undertaken in the United States since then.

The catastrophic accident of the Chernobyl reactor (northern Ukraine in the former Soviet Union) on April 26, 1986, enormously reinforced fears of nuclear power worldwide. This accident took place in a reactor without an outer protective shield, and it resulted from a thoughtless series of tests which could only be carried out (by theoreticians from Moscow?) by deliberately disarming the existing safety systems, apparently in ignorance of specific instabilities of reactors of the RBMK (Reaktor Bolshoi Moshchnosti Kanalni) type, originally designed to produce military plutonium. Victims of this accident include many of those who fought the fire, the "liquidators" involved in the following cleanup, and inhabitants of the nearby (quickly abandoned) town of Pripet (Pripyat) and surrounding areas. Although more than fifteen years have passed, assessment of the health effects remains uncertain and controversial.[27] Many people had to leave their homes, and farming was abandoned (at least as far as legally commercialized products are concerned) over a large area. It was only good luck that the wind was not blowing in the direction of Kiev that night. Only 100 km to the north, Chernobyl is close enough for day bus tours from that city of over 2,600,000 inhabitants. Since the accident, a "sarcophagus" was built to confine the highly radioactive products of the accident, although operation of the other reactors of the same complex continued for many years, the last one being shut down in December 2000. To be safe, a second sarcophagus needs to be built, with support to the impoverished Ukraine from the much wealthier European Union. After all, if the existing sarcophagus should collapse, the winds could blow radioactive dust to the west, as they did in part in 1986.

That story should not occult the serious damage caused by the military nuclear industry. On September 27, 1957, in the Mayak plant producing plutonium for Soviet weapons, part of the then-closed Chelyabinsk-40 complex located near the Ural Mountains, a storage vat exploded, spewing out enormous amounts of radioactive material, iodine-131, strontium-90, cesium-137, etc., all over the surrounding countryside, with heavy fallout in the waters of the Techa River (a tributary of the Ob, which flows to the Arctic Sea). Radioactivity due to that accident was detected in many parts of the world, in particular by the CIA, but the information was kept secret. Finally, in 1976, Soviet radiobiologist Jaurès Medvedev,[28] having succeeded in emigrating to England, revealed the disaster, only to have his information immediately denied by Sir John Hill, head of the U.K.'s atomic energy authority. What were the sources of Medvedev's information? For one thing, when authorities evacuate tens of thousands of people from large areas, forcing them to leave all their belongings behind, slaughtering and burying their livestock, razing their villages, and barring the roads, even a police state cannot stop news from getting out, and Medvedev collected and compared many stories. In addition, Medvedev carefully analyzed many research papers that appeared in specialized Soviet publications, describing results of purported hydrochemical and radioecological experiments in contaminated basins without naming the lakes, and furthermore he noted that the volumes involved were excessively large for "test basins." All his information has since been confirmed, in particular by declassification of CIA and Soviet archives.[29] Apart from this, much information has been released concerning the enormous radioactive contamination of many other sites for production of military fissile material, not only in the former USSR but also in the United States, in particular at Hanford (Washington), Rocky Flats (Colorado), and near Savannah (Georgia).

What about nuclear wastes, both civilian and military? The big problem is to keep them from getting into groundwater.[30] The most intensely radioactive isotopes soon disappear by their very radioactivity, and the longest-lived isotopes have lives longer than 10,000 years because their radioactivity is *not* intense. The most serious problems are with isotopes with half-lives from a few years to a few thousand years. In the United States, as in France, some specialists believe that satisfactory solutions already exist, and that one should get on with the job of putting the wastes into the designated repositories. The urgency of doing this is not so clear. The existing volume of nuclear wastes is still quite small. Why rush to put them underground? Why not leave them within the protected perimeters of the nuclear power plants (say next to the residences of the engineers in charge of safety?), for as long as it takes to convince citizens as well as scientists that a safe solution has indeed been found? Antinuclear activists retort that a satisfactory solution will never be found, and that allowing radioactive wastes to accumulate will only make it more and more difficult to phase out nuclear power. However, and we shall come back to this, nuclear power does not contribute to the emission of greenhouse gases, global warming, and the perturbation of the global water cycle.

PROBLEMS FOR THE 21ST CENTURY

A New Millennium?

What new millennium? True, the third millennium of the Christian calendar, adopted as the civil calendar in most countries, has begun.[1] And we are perhaps well into the eleventh millennium since the "agricultural revolution," when humankind began to change the landscape. About 15,000 years have gone by since the melting of the ice—in fact, not all, but only two-thirds of the ice. The Industrial Revolution began much less than 1,000 years ago, the sanitation and hygiene revolution only about a century ago. Among the most dramatic transformations of the last half century, the population "explosion" cannot but leave its mark on the future. From 1.6 billion in 1900, the world's population went to nearly 6 billion people in 2000. Growth was faster than exponential: about 0.7 percent per year from 1750 to 1950, but 1.9 percent per year between 1950 and 1975, adding billions of babies to the world's population. For the extra children to have any chance of survival, of reaching the age of one year, food production had to be increased by at least as much, and producing the extra food required withdrawing much more water from the ground and rivers.

What about the future? Will the 21st century see stabilization of world population?[2] Signs of this have already appeared: population growth has slowed since 1975, and for the first time in human history this resulted not from an increase in mortality, as when plague raged in the fourteenth century, but from reduction of the birthrate. In France, for reasons that still are not fully explained, this "demographic transition" had already begun 200 years ago; it spread to practically all European countries in the last decades of the twentieth century. The recent fall of the birthrate has accompanied and may be related to the increase in the standard of living and the achievement of some degree of women's rights and equality. Can stabilization and reduction of the birthrate be attained without this, and without authoritarian measures of the sort imposed in China? What about Third World nations? Continuation of the high 3 percent per year birthrate still observed in some countries is inconceivable. At that rate, population doubles in twenty-three years, with the number of children needing schooling increasing still faster. Under such conditions, who will teach them and give them the intellectual tools needed to survive in the 21st century, rather than indoctrination in suicidal fanaticism?

At the end of the last ice age, between 15,000 and 12,000 years ago, world population may have reached three million. Further increase only became possible with the development of farming, an act of appropriation of vegetation and of water on the land at the expense of other animal species. By the year 1 A.D., some 130 to 330 million human beings lived on the planet, and many of them could not have survived without irrigation. And since that time, humans have managed to nourish several billion people by increasing agricultural productivity and the area of land farmed. Will water availability be the final limit on further growth?

THE WATER RESOURCE

Let's take stock. First of all, the total amount of water on our planet stays pretty much constant; we can forget about loss to space, or gains from water-bearing comets. As for division of the Earth's enormous water stock between saltwater, ice, ground and surface freshwater, and water in the atmosphere (see table 6.1), the different shares vary fairly regularly with the seasons and they can also fluctuate from one year to another, but very little. Over thousands of years, however, much larger changes affect the division of water between ice and liquid. In any event, freshwater, essential for land vegetation, only exists as a temporary stage of the water cycle, between the sea and the sea, spending more or less time in that stage depending mostly on whether or not the "meteoric" water infiltrates underground (figs. 6.1 and 10.1). Apart from seafood, all our nourishment depends on freshwater falling from the skies onto the land. Three-fourths of the water used for farming returns directly to the atmosphere by evapotranspiration. Because of this, even though water is a renewable resource on the global scale, it hardly can be counted as renewable liquid water locally. By comparison, a factory usually returns (too often, unfortunately, in a polluted state) more than 90 percent of the water it withdraws from a river; indeed, many can operate with a practically closed water cycle and withdraw very little. Home water "consumption" also returns liquid wastewater. In rich advanced countries, efforts are made to treat wastewater and to return it clean to the rivers or the sea. But in many parts of the world, the wastewater, polluted by human and other wastes, reenters the surface or subsurface water without any treatment. Still, some of that pollution may act to fertilize the land. Wastewater carries with it health risks, but it can also be used as a resource for agriculture and aquaculture, depending on how it is managed; and sometimes it acts as both risk and resource.

How much water is needed, not just to quench thirst, but to feed and clothe a person? To produce the wheat for a kilogram (2.2 lbs.) of bread, a ton of water. For someone living on bread alone, with nutritious value of 2,000 Calories per day,[3] that corresponds to at least 200 cubic meters of water per year, assuming no losses between farmer and consumer. A more realistic figure is 400, thus more than 1,000 liters (a cubic meter or a ton) of water per day. Replacing bread by beef or mutton only one day out of ten more than doubles the effective water consumption. Figures given as the minimum for survival of an individual range from 500 to 1,000 cubic meters of water per year (i.e., 1,300 to 2,700 liters per day). In 1990, 125 million human beings had access to less than 1,000 cubic meters per year. In the wealthy countries of America and Europe, people currently use directly or indirectly close to 2,000 cubic meters per year, in some cases even much more.

How does rainfall satisfy these needs? In well-watered regions—for example, much of France and the eastern United States—most agricultural production does not need irrigation. In other areas, where farmers would like in times of drought to be able to tap stored water surpluses, irrigation doesn't exist. Much of the rainwater that falls on the fields or the woods returns to the atmosphere by evapotranspiration. Still, over much of the world, the largest share of the water used to satisfy human needs—for irrigated farming over 16 percent of the world's land surface, for home consumption, for cooling of power plants, for industry, etc.—comes from withdrawal of runoff or groundwater. We can estimate the renewable water resource in two different ways: by considering total rainfall over land, 110 trillion cubic meters per year; alternatively, considering only runoff, more than 40 trillion cubic meters per year (see table 12.1 and fig. 6.1).

Today, by way of agriculture, livestock, and forestry, humans appropriate more than a fourth of the world biosphere's "net primary production," the production of living plant matter by photosynthesis.[4] Out of the 70 trillion cubic meters of precipitation that fall on the land but return to the atmosphere by evapotranspiration, about 28 trillion pass by way of land exploited by humans. Whether we measure our activities on Earth by the areas that we occupy and manage more or less well, by the plant matter that we consume (directly, or by way of eating meat), or by the rainwater that we use mainly by the intermediary of plants, our impacts on Mother Nature are far from small, even on the scale of the planet as a whole. As deforestation advances in the Tropics, how will this expansion of human action affect the division of the waters? We don't know. Where pasture, food crops, or tree farming take the place of the tropical forest (assuming they

TABLE 12.1 RUNOFF:
RENEWABLE FRESHWATER FLUXES AND WATER AVAILABLE PER PERSON

Annual runoff figures for the world and for each region are given in column 2 in cubic kilometers per year, i.e., billions of cubic meters per year. Per capita runoff figures (columns 4 and 5) are in cubic meters per year per person.

Region	Annual Runoff (cu. km per yr.)	Share (%)	Annual Runoff per Capita	"Accessible" Freshwater
World	**40,700**	**100**	**7,500**	**6,100**
Europe	3,240	8	4,600	3,750
Asia	14,550	36	4,400	4,200
Africa	4,320	11	6,400	5,300
North America, Central America	6,200	15	14,300	11,700
South America	10,420	25	35,000	16,800
Australia, Oceania	1,970	5	74,000	40,000

don't quickly become barren wasteland), will runoff increase at the expense of evapotranspiration, or vice versa?

Because rainfall over some areas (in particular the interior of the Amazon River basin) consists mostly of regionally recycled water from evapotranspiration by forest vegetation, the question has been raised whether deforestation will not lead to the drying up of those areas. On average over the globe, as noted earlier (fig. 6.1), nearly two-thirds of the rain that falls on land returns to the atmosphere as evapotranspiration rather than running off to the sea in streams and rivers. In some areas, in particular on sloping terrain, deforestation reinforces runoff at the expense of evapotranspiration, aggravating soil erosion at the same time. However, the reverse may hold in other areas, i.e., runoff may be reduced and evapotranspiration enhanced. That depends on the vegetation replacing the forest trees, the properties of the soil, and the skill with which the land is managed. In any case, extension of agriculture and ranching to previously wild areas inevitably encroaches on the living space of many wild species, considered by ranchers and farmer as useless if not as pests to be exterminated. Such expansion of the human-controlled domain thus reduces what is called *biodiversity*, the enormous variety of life forms on the planet. Moreover, as on many occasions in the past, such expansion diminishes the chances of cultural or even of physical survival of those peoples that still practice hunting and gathering on lands that no one coveted before. Considerable difficulties stand in the way of attempts to protect the tropical forest (in the Amazon, in Africa, in Indonesia), even when the arguments given in favor of such protection are couched in terms of protection of economic interests of modern society. Deforestation does *not* constitute a threat to the oxygen in the atmosphere, despite the myth or metaphor that calls the Amazon forest the "lungs of the Earth." Maybe deforestation leads to climate and hydrological catastrophes, but maybe not. What then? Should all attempts to protect the Amazon forest and its native inhabitants then be abandoned? It seems to me that there is a sort of inverse hypocrisy when environmentalists attempt to justify moral reactions and basic decency by purely egotistical arguments of economics.[5]

Table 12.1 includes a column for "accessible" freshwater. The freshwater flow *not* exploited and considered inaccessible includes practically all the upper Amazon basin's flow (more than 5,000 cubic km per year), Eurasian and Canadian rivers flowing to the Arctic Ocean, and about half of the

Congo River's flow (fig. 10.3). At present, very few people live in these regions through which flows a significant fraction of the world's renewable freshwater. Asia has enormous freshwater resources but, except in Siberia, intensive farming has long exploited them to feed an ever-increasing population more and more efficiently, as time has passed. In Europe, such development and population increase took place more recently, over an area and with water resources half as large. In both cases, the per capita accessible water resource amounts to about 4,000 cubic meters per year, enough to support the present large populations (Asia: 61% of the world's population; Europe: 13%), considering the development of agricultural productivity. In Africa, the per capita water resource considered accessible today is only slightly larger. In the "new" worlds of the Americas and Oceania, farmer and rancher colonists from Europe displaced, infected, and often massacred the indigenous populations, sometimes replacing them with slaves from Africa. Because population densities have remained low compared with Old World norms, present New World inhabitants enjoy at least twice as large per capita available freshwater. About 40 percent of the world's runoff is found in the Americas for only 14 percent of the world's population. In Australia, although water is scarce, humans are even scarcer, so that the per capita water resource is ample. And so long as the waters of the jungle and central highlands of New Guinea are considered "inaccessible," their inhabitants will be left alone!

SATISFACTION OF WATER NEEDS

With accessible water resources of at least 3,500 cubic meters per person per year in different regions of the world, compared with an absolute vital minimum of 500 cubic meters per person per year, and 2,000 corresponding to relative comfort, the margin between needs and resources may appear quite safe, especially considering evaporated and transpired rainwater in addition to runoff. However, nothing in nature guarantees that the water resource, specifically the renewable freshwater flux, automatically keeps up with human population growth. If population doubles one more time, in Europe, Asia, or Africa, the per capita water resource may well fall below the 2,000 level. Of course, in Asia or in Africa, plenty of people would be happy to be sure of even 1,000 cubic meters per year, but in rich countries of Europe, as in North America, the 2,000 level is considered just comfortable. At present, 30 percent of runoff in Europe is already being exploited; the average for the globe is 20 percent. Despite the apparent wildness of vast areas of the world, humankind plays a major role in the water cycle, withdrawing, evaporating, consuming, soiling, and filtering water, from its flow over the land and through vegetation. However, despite worldwide trade in bottled mineral waters, despite projects to ship freshwater from the Great Lakes to the Persian Gulf or from Turkey to Israel, such large-scale freshwater imports remain a privilege of the rich who can exchange oil or high-tech products for water. During coming decades, will human ingenuity find ways to increase the water resource, not only by tapping presently "inaccessible" water but also by accelerating the water cycle? Such increase is not inconceivable, but such projects will require still greater interference in the processes of nature, still more marked humanization (or industrialization) of the environment. Instead of increasing supply, can one limit demand by limiting population and by enhancing the efficiency of water use to satisfy human needs? There again, enhancing efficiency may involve changing the landscape, replacing forests by farms, putting land "to use." That may also mean building more dams and reservoirs to irrigate dry lands.

Distribution of water resources is uneven between continents, extremely uneven between different countries and regions (table 12.2). Africa includes the moist tropical zone (for example, Gabon, with more than 100,000 cubic meters per person per year), the almost perfectly dry Sahara Desert (where very few people live, but caravans pass), and North Africa (e.g., Algeria, less than 1,000 cubic meters per person per year). The United States enjoys nearly 10,000 cubic meters per person per year, but the figure is much lower for Arizona, much higher for Alaska! In Europe, Spain's per capita resource (3,000) is only one-seventh that of Sweden (21,000); but, surprisingly, water available per person is still scarcer in Belgium, Poland, and the Ukraine, although those countries are wetter than Spain.

How much of the available water resource is actually used? In Europe, runoff exploitation ranges from 5 percent (Hungary) to 50 percent (Bulgaria, the Ukraine), with most other countries in between, but it has already reached 70 percent in Belgium. Some areas have far more freshwater than their human inhabitants need (Alaska, Gabon, French Guiana, Iceland, Norway, and others), with a minuscule exploitation rate. But in other areas, ex-

TABLE 12.2 RUNOFF BY REGION:
AVAILABLE FRESHWATER PER PERSON AND EXPLOITATION RATE

Region/country	Cubic Meters per Person per Year in 1990	Exploitation Rate (%)
Algeria	550	30
Belgium	1,500	70
Brazil	44,000	less than 1
China	3,400	25
France	3,300	15
Gabon	116,000	less than 1
Germany	2,300	30
Iceland	667,000	less than 1
India	2,200	20
Iraq	6,500	55
Israel	460	100
Ivory Coast	7,400	1
Japan	4,300	20
Libya	150	600
Portugal	6,500	10
Russia	28,000	3
Sweden	21,000	1
United Arab Emirates	380	150
U.S.A.	9,900	20
but in Alaska	more than 100,000	less than 1

ploitation exceeds 100 percent—in the Arabian peninsula, Israel, Libya, some areas of the southwestern United States, and some enclaves such as Djibouti. Saudi Arabia consumes freshwater at ten times the rate of renewal. How is an exploitation rate greater than 100 percent possible, and what does it mean? Some rich cities at the edge of the desert along the Persian Gulf get their drinking water by desalination of seawater, perfectly feasible but expensive at about a dollar per cubic meter. However, many places with exploitation rates above 100 percent are in fact tapping a nonrenewable resource, just as in mining: pumping groundwater at a rate far in excess of the rate of renewal of the water table by infiltration. Often, this "fossil" water corresponds to a bygone climate that reigned and rained thousands of years ago. Such resources are used to water crops in the desert, in the Sahara, Saudi Arabia, Australia, and Arizona. On the relatively dry plain of Nebraska and eastern Colorado, pumping of the Ogallala aquifer will probably dry it up completely within a couple of decades.

WATER MANAGEMENT: COOPERATION OR CONFLICT?

Although population growth is slowing down, it won't stop short, and it raises acute problems for water management. First of all, especially in those countries where population continues to rise, increasing food production necessarily requires additional water and fertilizer, and even if enough water exists, indiscriminate use of fertilizer and pesticides may pollute it. Second, since much of the population increase takes place in rapidly growing monster cities, how can enough safe drinking water be supplied to them, and how can the growing flux of waste be cleaned up? Even today, millions of people in the poor countries have no access to safe drinking water, and sewage flows in what passes for the street in thousands of shantytowns.

It's often said that water is part of a common human heritage, but in what sense can this be true? Water problems must first be solved at the local level. Rainfall on a field appears first of all as a gift from heaven, from God if you will, but His ways are not so simple, as we shall see.[6] As for water drawn from the village well, used by villagers to drink, to give drink to their animals, to cook, and to wash, that water is first of all part of the village heritage, and it's up to the villagers to dig the wells, to maintain them, and to protect them from pollution. The farmer who digs a well or who uses water from a spring on his own property (in many lands having his wife carry it) may consider that it's his own water. However, when users realize that their well or spring depends on a water table also tapped by other users of other wells and springs located in the next township, and that one or the other or all of the wells may dry up, depending on who uses how much water, they no longer can settle their water problems on a purely local level. That is even more true for use of river water, where upstream withdrawal and waste discharge by one user affect all downstream users. Water management is perhaps best organized at the scale of the basin, where different users and stakeholders in competition—farmers, ranchers, miners, factory operators, cities, power plants, tourists—have somehow to reconcile their competing and at times contradictory interests. Some of them would like to get rid of their wastes at the lowest cost possible, others insist on their rights to "pure" water. Different societies have developed different methods for dealing with these problems, through bodies of law and tradition, through different institutions. Does a mining operation have the right to use and pollute all the water of a stream running through its mountain land, when in the plains, downstream, ranchers need the water for cattle, towns for their citizens? Different states give different answers, and often no solution can be found in local jurisdictions.[7] The word *rival* comes from the Latin *rivalis*, a person using the same stream as another.

Even in the United States, with over 200 years of democratically elected federal government, centralized authority is often viewed with suspicion although federal funds are always welcome. The same holds in the young European Union, where politicians tend to leave difficult and unpopular decisions to the appointed delegates of the European Commission in Brussels, which everyone is happy to complain about. Europe has developed the doctrine of "subsidiarity," that political decisions should be left to the smallest political unit that is competent to make the decision. How can this be applied to water management? What is the appropriate scale? Although the European Commission holds no brief to regulate every watershed in France and Spain, the way water is used in Aquitaine (a region of southwest France) can affect fishing in the Bay of Biscay, a matter of interest to Spain and other European countries as well as to France. The "stakeholders" in a river basin therefore include those whose activities depend on the quality and amount of water that the river sends to the sea. For rivers such as the Rhine or the Danube, flowing through several different European countries (fig. 10.3), basin management must be on the international scale, and indeed such international river management agencies have long been in existence in Europe and North America.[8] Others have begun to take shape in Asia. Rivals India and Pakistan signed the Indus Waters Treaty in September 1960, and they have managed these waters in a spirit of cooperation despite the ongoing conflict over Kashmir. On the other hand, although a four-country (Kampuchea, Laos, Thailand, and Vietnam) framework for cooperative development of the Mekong River basin was established as early as 1957 with United Nations and World Bank support, the Vietnam

War and later the Vietnamese invasion of Cambodia (Kampuchea) limited what could be accomplished.[9] Resolution of conflicts of interest in water remains difficult, even for countries or states otherwise at peace. Hungary has sought to block Slovakia's completion of the big Danube dam project started when the two countries were part of the "People's Democratic Republics." In North America, the waters of the Colorado River, born on the Western Slope of the Rockies, flow through desert and semidesert areas of five U.S. states, passing through Mexico before reaching the Gulf of California. All compete and litigate for use of this water. The Los Angeles metropolitan area is always thirsty, as is Phoenix, Arizona. And in California's Palm Springs, winter quarters of Hollywood stars and other multimillion-aires, the lawns and golf courses are always green, thanks to diverted water that evaporates into the dry Mojave Desert air, cooling the area.

The giant James Bay hydroelectric complex supplies the Province of Quebec with abundant electricity (more than 10,000 megawatts), but it has disrupted the ancestral way of life of the indigenous Cris and Inuit inhabitants of northern Quebec province. During the Soviet era, some visionaries proposed to irrigate "new lands" in Central Asia by reversing the course of some of the big rivers flowing northward to the Arctic, but today, after the disaster of the drying up of the Aral Sea, and with growing environmental awareness in the Russian public, few want to translate those dreams into what many suspect would be a nightmarish reality. Every so often, the idea of a titanic transport scheme for bringing water from the American and Canadian Northwest to southern California rears its head, but it's probably another Hollywood fiction. Indulgence of L.A.'s never-ending thirst has its limits. Generally speaking, the big water problems will have to be resolved at the river basin level, and not in trying to transfer enormous quantities of water from one basin to another.[10] Even within a basin, complications abound. Some would say that the technical solutions exist, and that all we need is the political will and courage.[11] However, a good dose of diplomacy will help. And we still need to improve our understanding of land and water processes, and of people's relation to them, in order to maintain a supply of daily bread and water to the six billion human beings (soon eight or even twelve) who depend on the uneven distribution of the water resources on our planet.

In its heyday, the USSR, like Pharaoh's Egypt, organized colossal construction projects, building giant dams and hydroelectric power plants, flooding and irrigating large land areas. The transformed river basins stretch across what are now several different independent countries—Belarus, the Ukraine, the Russian Federation, and the former Soviet republics of Central Asia—with nuclear power plants located here and there. Today, few defend the way the waters of Lake Sevan (Armenia) were used, or the diversion of the rivers that flowed into the Aral Sea. As noted earlier, neither can the Soviet-inspired Aswân High Dam on the Nile in Egypt be regarded as a total success. At present, however, only Egypt makes intensive use of Nile water, which comes from Ethiopia and central Africa by way of Sudan. But suppose that one day Ethiopia decides to reduce soil erosion and to produce power by building a big dam on the Blue Nile?[12]

Such a question, hypothetical (so far) in the case of the Nile, has become a very real one in the case of Turkey's Atatürk Dam and more generally its southeast Anatolia project or GAP, controlling and harnessing the headwaters of the Tigris and the Euphrates in the Taurus Mountains.[13] This certainly has the potential of reducing downstream water flow vital for the economies of Syria and Iraq (fig. 12.1). In addition, if indeed Turkey develops irrigated agriculture while protecting its watershed from erosion, downstream water will contain less suspended matter, but possibly more pesticides, fertilizer, and salt. Turkey has asserted its desire to reach cooperative agreements for "equitable, reasonable, and optimal utilization" of Tigris and Euphrates waters, proposing to supply electric power as well as potable water to downstream Syria and Iraq, and also to pipe water to Jordan and Saudi Arabia. Note that construction of a further branch to Israel would be technically quite easy, but politically inconceivable without a final comprehensive Israeli-Arab peace. The Tigris and Euphrates Rivers, two of the four rivers leaving the Garden of Eden (Genesis 2:14), have always played an essential role in the irrigation of food-producing lands from the times of Mesopotamia, Assyria, and Babylon to modern-day Syria and Iraq, including those regions with large Kurdish populations. "Hydropolitics" was perhaps not an important factor in 1920, when a Kurdish state was promised (only to be abandoned later), but with its enormous investments in the GAP, resistance of increasingly powerful Turkey to any hint of Kurdish autonomy, not to mention statehood, is stronger than ever.[14] Although the growing importance of the water dimension in Middle Eastern politics cannot be denied, it must be recognized that both the Iran-Iraq war and the 1991 Gulf War were still

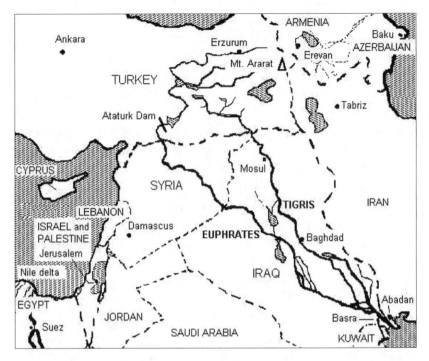

FIGURE 12.1 The Tigris and Euphrates Rivers originate in southeast Anatolia (Turkey) and now run through Syria and Iraq, finally emptying together into the Persian Gulf by way of the Shatt-el-Arab channel between Basra (Iraq) and Abadan (Iran). The Ataturk Dam is the largest of the many dams constructed or planned in Turkey's GAP.

about oil and access to the sea rather than freshwater. Also, despite some scary suggestions, the possibility of stopping water from reaching Iraq was not put into effect as a weapon in 1991.

WATER BETWEEN ISRAEL AND THE ARABS

Sharp competition surrounds the much smaller water flows from Jordan, Syria, and Lebanon into the Jordan River of Israel, Jordan, and the Palestinian authority, but does the water dimension play a major role in the conflict between Israel and the Arabs? It seems to me that the answer must be both yes and no. The basic conflict is about land but, over most of human history, that meant land with water. Only recently has anyone coveted desert, and only oil-rich desert at that.[15] When the seminomadic Israelite tribe of Abram left Babylon, and Abram become Abraham profited from a temporary decline in Egypt's power to invade and occupy, with his sons Ishmael and Isaac, the land of Canaan, the problem was indeed to find water (Genesis 21:19–31 and 26:15–23), which they did, at Beersheba.[16] Drought forced emigration to Egypt with its reliable Nile water, stockpiled grain, and slavery, but after the Exodus and return to the Promised Land,[17] and following the establishment of the united kingdom of Israel, the choice of Jerusalem as capital by King David (1004–965 B.C.) was fortunate in that the city had good access to water in the nearby pulsating karstic spring of En-Gihon. Between 705 and 701 B.C., Hezekiah, king of Judah, anticipating an Assyrian invasion, had a remarkable underground tunnel dug through over 500 meters of solid rock to convey water from the Gihon spring (which he had hid by surrounding walls) to the Siloam Pool.[18] This was an essential factor in the city's "miraculous" resistance to the siege by Sennacherib.[19] It did not, however, prevent the fall of Jerusalem to the Babylonians in 586 B.C., when the First Temple was destroyed. Following the return of the Jews from Babylonian exile in 538 B.C. and establishment of the Second Temple, the water supply was at times placed under the responsibility of the High Priest; but by the time of Jesus, it had been forgotten that the waters in the Siloam Pool actually came from Gihon spring.[20] Jerusalem, stronghold of the Great Revolt against the Roman Empire, was destroyed in 70 A.D. by Titus's legions, who had starved out the resistance by surrounding the city with walls.

Following several centuries of Roman and, later, Byzantine (Christian) rule, following the Muslim conquest in 638 A.D. and Arab rule briefly interrupted by the Crusader incursion (1009 to 1187 A.D.), the Ottoman Empire ruled Palestine as a minor province, by way of governors in Damascus, from 1516 up to World War I. Modern Zionism appeared only in the nineteenth century, a reaction against European anti-Semitism as much as a semireligious movement, with an ideology of return to the land, land presumably with water. When the British mandate was established in 1919, conditions in Palestine and indeed over much of North Africa and the Near East had become very different from Roman times when these areas produced large amounts of grain. Although some tended to put the blame on the grazing by goats belonging to Bedouins, on Arabs in general, or on Ottoman neglect, the desertification of many of these regions may also have been related to climate change. At any rate, the Zionist ambition was to "make the desert bloom," and the ideology included many elements in common both with American "can-do" philosophy and with technological optimism of the socialist movement. The Israeli-Arab conflict, a competition for land, necessarily also has a water dimension, because application of modern agricultural methods cannot do away with the need for water even if it can in principle render water use more efficient. With the development of the largely modern and rich Israeli society, water withdrawals have grown more rapidly than the population, and Israel now uses at least 100 percent of its renewable freshwater resources. Growth of the Palestinian Arab population, and significant although still limited economic development, also have increased water demand.

"Hydropolitics" has become distinct from geopolitics both because of increased awareness of the interconnectedness of ground and surface water resources over extended areas, and because it is now technically possible both to stop river flow and to transport water over large distances. In the Middle East, competition and conflict between Assyria and Babylon once concerned living space along the Tigris and Euphrates Rivers, but today Syria and Iraq fear Turkey's ability to impound the water before it reaches them. In North Africa, Libya's "Great Man-Made River" may be pumping some water from an aquifer extending in part under Egypt. In the Near East, Palestine's strategic position between Egypt (the Nile) and Mesopotamia (the Tigris and Euphrates) inevitably made it a focus of contention in biblical times. Today, in addition to the Israeli-Arab conflict over land, there is competition and potential conflict over water. Israel's *National Water Carrier*, completed in 1964 (three years before the Six-Day War), pumps more than 400 million cubic meters of water per year from the Sea of Galilee (Kinneret), supplying farms as far as 250 km (150 mi.) to the south (fig. 12.2). However, although Kinneret lies entirely within Israel,[21] less than half of the upper Jordan River flow to Kinneret originates in Israel, essentially the 250 million cubic meters per year from the spring of Dan, the rest coming from Lebanon and Syria.[22] Although Israel gained control of some of the Jordan's tributaries in the Six-Day War of June 1967, this was hardly a factor in that conflict, which was triggered when Egypt took measures to block Israel's Red Sea outlet to the south. Was it a significant factor in 1979–1982 Israeli operations in Lebanese territory, which reached the Litani River? The answer is not clear.[23] However, Israel is extremely sensitive to any Syrian or Lebanese attempts to withdraw water from the Banias or Hasbani Rivers, because of the potential impact of such withdrawals on the Jordan River's flow into Kinneret.[24] Israeli military action in 1965 and 1966 effectively ended Syria's Headwater Diversion Plan, and following the Six-Day War, Israel controlled nearly all of the Jordan headwaters.

Serious conflict also exists over groundwater, both directly and indirectly related to the conflict over land. The coastal plain aquifer extends from Carmel (near Haifa) in the north to the Palestinian Gaza Strip in the south (fig. 12.2), and although enough rain falls in winter to support some farming, intensive modern Israeli agriculture requires pumping of groundwater. In addition, because of extremely high population density in the Gaza Strip, as well as very high water usage by Jewish settler farms, the aquifer there is being pumped at about 120 million cubic meters per year, nearly twice the sustainable rate, and salt content has reached excessively high levels. Over the Israeli portion of the coastal plain, pumping of groundwater exceeds recharge of the aquifer (infiltration of rainfall together with recharge from irrigated lands and wastewater recycling) by some 240 to 300 million cubic meters per year, there too compounding the risk of salinization by seawater intrusion.

Further complicating this issue, recharge of the mostly Israeli coastal aquifer depends in part on infiltration from the fairly well-watered mountain aquifer, over essentially Palestinian territory, historically corresponding to the post-Solomon pre-Roman Jewish kingdoms of Judah and Israel, areas called Judea and Samaria by Israelis. Indeed, some charge that as much as a third of the water used in Israel comes from groundwater from rain that falls over the western mountain aquifer and infiltrates westward toward the coastal aquifer. Until the Israeli occupation's start in 1967, the inhabitants of this "West Bank" of Jordan practiced mostly rainfed agriculture, with some groundwater withdrawal for urban water supply. With population growth and introduction of more intensive agriculture, increased withdrawals from the western mountain aquifer reduce the water available in the coastal plain below and may allow enhanced seawater intrusion into the coastal aquifer. Israeli specialists also fear pollution flowing into the coastal aquifer if West Bank agriculture intensifies, and recognizing that some of the water pumped in the coastal plain comes from the mountain aquifer, they argue that prior use (prior to the Six-Day

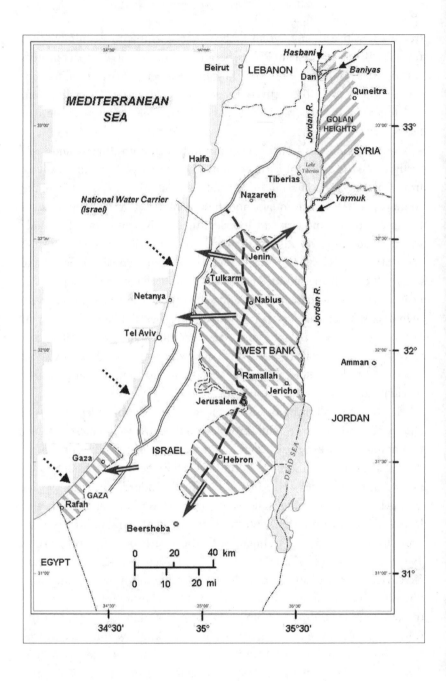

War) gives them a legal voice on what is done with and to the mountain water. The argument has its analogy in Syrian and Iraqi objections to Turkey's GAP. Israeli geohydrologist Arie Issar has, however, argued that the one dollar per cubic meter cost of desalinating Mediterranean water could easily be borne by the rich Israeli economy, and that (writing in an Israeli voice) "we and the Palestinians have the obligation to draw up a co-operative regional plan for water resources utilization (as part of a regional plan with our other neighbors), that will permit development of our respective agricultural and urban sectors. In their well-being lies our own, while their economic deterioration, with its great attendant disappointment in the peace process, is a sure recipe for the ascendancy of religious fundamentalism and the continued reign of terror in the region."[25] Palestinians and some Israelis accuse the Israeli occupation authorities of unfairly limiting access to water by Arab inhabitants of the West Bank, and in some cases of including previously operating wells in forbidden security areas, while giving a much freer rein to new Jewish settlements with high water-consumption rates. Israel has also been accused of "stealing" rain that would normally fall on Palestinian lands by cloud-seeding operations in northern Israel.[26] Moreover, some Palestinians believe that increased salinity of water from wells supplying Jericho results from intrusion of Dead Sea water due to pumping of deeper groundwater by Jewish settlements in the area of the eastern mountain aquifer. Thus, the conflict

---

FIGURE 12.2 (*opposite page*) Water between Israel and its neighbors. The heavy dashed line represents the groundwater divide between west and east mountain aquifers, running mostly through Palestinian territory in the "West Bank," occupied by (formerly Trans-)Jordan from 1948 to 1967, by Israel since June 1967. The heavy arrows represent groundwater flow, mostly toward Israel. Dashed arrows represent saltwater intrusion along the coastal aquifer from south of Haifa to the Gaza Strip (Palestinian territory, occupied by Egypt from 1948 to 1967, by Israel since June 1967). In the northeast, occupation of Syrian territory since June 1967 has given Israel control over most of the Jordan River's headwaters. Boundaries on the map correspond to the 1949 Armistice Demarcation Line and the boundary of the former Palestine Mandate. Jewish settlements within Palestinian and Syrian territories are not shown. Only the main components of Israel's National Water Carrier are shown. This map is adapted from United Nations map no. 3652 (September 1991), which includes the sentence: "The designation employed and the presentation of material on this map do not imply the expression of any opinion whatsoever on the part of the Secretariat of the United Nations concerning the legal status of any country, territory, city or area or of its authorities or concerning the delimitation of its frontiers or boundaries." The Sea of Galilee (Kinneret) is identified as Lake Tiberias.

over the land occupied by Jewish settlements in the West Bank concerns not just that land but an interconnected system of runoff and infiltration of water.

Despite all these conflicts, there was progress toward peace in the 1990s. The Israel-Jordan peace treaty of 1994 specifically includes water-sharing issues, as does the 1993 Israeli-Palestinian "Declaration of Principles on Interim Self-government Arrangements," with its call for creation of a Palestinian Water Authority. Israeli and Arab specialists have met many times and worked together on issues of water management and the reduction of water pollution, regarding both groundwater and runoff in the Near East and discharges into the Mediterranean. All of this now (in mid-2002) appears severely compromised, as religious and nationalist fanaticism, fear, and violence gain the upper hand in an ever-worsening situation.[27] Because the atmosphere takes and brings water over borders across the entire world, because water on and in the ground inevitably links the destinies of all those who inhabit a river basin or even larger region, it appears difficult to solve the increasingly serious water crisis facing both the rich and scientifically advanced Israelis and the impoverished Palestinians without cooperation. But what are the chances of cooperation, if each side is blind to the legitimacy of the other's presence?

CHAPTER 13

BUTTERFLIES AND HUMANS IN
A WARMING GREENHOUSE

THE BUTTERFLY EFFECT

Some self-styled ecologists regret that we humans are no longer just one species of mammal among many others, carrying on our existence in a cozy well-defined ecological niche in the midst of the splendid balance of nature. But there never was such a golden age of equilibrium, and we know today that all nature is change. We know also that our acts as humans have consequences that escape our control, and that our artificial creations of today become new phenomena of nature that in part determine what tomorrow will bring. The newcomers who plowed the dry lands of the southern Great Plains, following the 1889 land rush opening the Indian Territory, had little inkling that the topsoil would blow away in the terrible Oklahoma dust storms of the 1930s. As Woody Guthrie put it: "This dusty old dust is getting my home, and I've got to be moving along." And many of those who took off for the Californian Eden only got to harvest grapes of wrath. In another, more recent register, in 1938 Du Pont de Nemours invented the chlorofluorocarbons (CFCs), which for almost forty years appeared to be veritable miracle molecules, risk-free and extraordinarily useful in many applications from refrigeration to spray cans. But since 1974, we know that their residence in the atmosphere only ends when they destroy ozone in the stratosphere; since 1987, manufacture of CFCs has been terminated.[1] But replacing them is no simple matter, and the pros and cons of each new product must be weighed.

Can we foresee the future? Knowing enough of the present state of things, can we compute the future course of events? That is, after all, what weather prediction tries to do. Every day, at least twice a day, from ground stations, weather balloons, and satellites, meteorologists collect measurements of the state of the atmosphere, in particular of the variations of temperature, humidity, and wind with altitude and pressure. Using Newton's laws together with the laws of thermodynamics and the principles of conservation of energy and of mass, we can write the equations that govern the winds, the exchange of energy between air, sea, land, and ice, and the changes of state of water. With modern computers, the equations can be solved, and the computations give fairly good forecasts—certainly better than they're given credit for—12, 24, 48, and even 72 hours in advance. Beyond that, weather forecasts often go wrong, and forecasters readily admit it.

Do the errors arise only from insufficiently accurate measurements, from errors in the equations, or from the approximations made in the calculations? It seems not. Early in the nineteenth century, Laplace wrote that if some superhuman intelligence knew the exact positions and velocities of all particles (we might write atoms) at one instant in time, all their movements and every event in the future would be determined and could be calculated until the end of time.[2] We know today that physics forbids such perfect knowledge. Some apparently insignificant details always remain unknown but end up totally changing the course of events, a phenomenon described by modern "chaos" theory. In the atmosphere, we certainly have no hope of knowing everything at a given time, no way to follow every cubic meter of air, no chance of predicting in detail the birth of a small instability in the circulation of the atmosphere, or the growth of a tiny eddy that develops into a cyclone by the end of the week while another one disappears in a few minutes. Because of these factors, deterministic Laplace-style weather prediction often goes wrong in a few days, and there appears to be no hope of extending its range beyond two weeks. This limitation on predictability has been called the "butterfly effect." Imagine a butterfly flapping its wings in Brazil, or a little girl blowing out her birthday candles in a garden in Orlando, setting into motion a minuscule puff of air, which from puff to eddy and still more eddies, from evaporation to condensation to clouds, finally triggers a storm that sweeps across Cape Cod or crosses the Atlantic to England.

With "butterflies" in the computer, or in the crystal ball, is there no hope of seeing beyond the end of our nose, our headlights, or next week? It's not that bad! Weather prediction models do indeed give forecasts that stray from reality after several days, that send a snowstorm to Norfolk rather than to Roanoke, that put clouds into Boston's sky two days too soon, when they're passing over Cleveland. But the operational weather services continually correct the predictions, recalling the models to reality by injecting new data every twelve hours as the situation changes. But how can that help when they want to predict climate? By making sure that the models strictly enforce the laws of conservation of energy and of conservation of matter, by taking particular care of conservation of water (keeping count of its solid, liquid, and gaseous states), relatively realistic "simulations" of climate can indeed be obtained from long "runs" of the models.

Model calculations made, say, in June 2001 can simulate January climate by computing thirty different fictitious but realistic weather histories for January, and these "virtual Januarys" compare well enough overall with months of January actually observed over the past thirty years. However, these calculations certainly cannot tell us whether or not it will snow in Washington on Inauguration Day in 2005 or 2009, although they do provide a basis for calculating the odds so long as climate itself does not change. Climate-modeling claims only to determine the statistical properties of the weather, i.e., averages and variability of temperatures, precipitation, and other meteorological quantities.

Nevertheless, many weather services, in particular the U.S. National Weather Service, have taken the risk of making forecasts several months in advance, not of day-to-day weather but rather of the characteristics of a season, whether they expect it to be colder or warmer, wetter or dryer than average, in different regions of the country. As early as spring 1997, meteorologists predicted the strong El Niño event (see fig. 7.3) from October 1997 to March 1998, with rain in Peru, drought in Indonesia, and strong storms reaching California. The basis for such a verifiable prediction resides in the fact that the distribution of water temperature at and near the surface of the Pacific at a particular date strongly influences weather all around the Pacific for a few months, that the models now correctly represent that influence, and that timely data exist. The international scientific community had indeed set up a system of drifting buoys to measure temperatures in the tropical Pacific. That does not contradict the butterfly effect. The models did predict correctly, as early as April 1997, that California would bear the brunt of many storms, but they could not forecast more than a few days in advance when and where on the coast those storms would strike. Similarly, the mild 1997–98 winter of the northeast United States and eastern Canada was no surprise, but only a few hours' advance notice were possible for the terrible ice storm that paralyzed the province of Quebec during that winter.

The Greenhouse Effect

The natural greenhouse effect (see chapter 6) results from absorption of the Earth's thermal infrared radiation by certain gases of the atmosphere, the most important of these being water vapor (H_2O). Only 0.001 percent of the planet's water exists in this state, and water vapor constitutes only 0.25 percent of the mass of the atmosphere. Nevertheless, water vapor plays the principal role both for absorption of infrared radiation and for other exchanges of heat within the atmosphere and between the atmosphere and the surface. Water vapor is the most important of the greenhouse gases, but the contribution of the next most important, carbon dioxide (CO_2), cannot be neglected. Like H_2O, CO_2 is relatively rare in the atmosphere, constituting only 0.04 percent of its mass. The amount of CO_2 in the atmosphere has varied enormously in the course of time, with relatively high values in the early ages of the Earth, decreasing later. Over the last two million years at least, the concentration of CO_2 in the atmosphere has followed the changes in the shares of ice and liquid water on our planet. Analysis of ancient trapped air bubbles in the depths of the ice caps of Antarctica and Greenland has shown that the concentration of CO_2 fell to 210 (or perhaps as low as 180) parts per million in units of volume (ppm/v or ppm) at the time of the Last Glacial Maximum 18,000 years ago; but as the ice retreated, the abundance of CO_2 rose, reaching 270 ppm at the start of the Industrial Revolution 250 years ago. Since 1750, and especially since 1950, burning of fossil fuels (coal, oil) has increased the CO_2 level further, to 315 ppm in 1958, 335 in 1975, over 370 ppm in 2000. These increases are comparable in size to the natural increase due to the end of the great ice age, but they create conditions not seen in the atmosphere for millions of years at least. Today, for carbon dioxide in the atmosphere, for water flowing on the land, for the makeup of the biosphere, and for the global nitrogen cycle massively modified by chemical fertilizer use, humankind acts at the center of the planetary stage.

The best hope for completion of the demographic transition and stabilization of world population lies in reducing poverty in the Third World, but development depends in part on using energy, energy up to now mostly produced by burning fossil fuels. China, the largest of the developing countries, possesses enormous reserves of coal, as well as a huge and still only partly tapped potential in hydroelectric power. Even in Western Europe, several countries (for example, Greece and Portugal) still have far to go before reaching the living standard of their neighbors, and a high standard of living seems to require high energy consumption. In many already rich countries, the owners of coal mines or oil wells would like to continue to dig, drill, sell, and export. How high will the carbon dioxide concentration go? In the present state of affairs,[3] a level of 450 ppm appears practically inevitable before the end of the 21st century, and global adoption of the Bush-Cheney waste-and-burn energy policy would lead to levels well above 600 ppm. In addition to carbon dioxide produced by fossil fuel burning, other gases (methane, CFCs, nitrous oxide, etc.), further intensifying the greenhouse effect, are being added to the atmosphere by human activities. Methane (CH_4) production by certain bacteria living in oxygen-free environments (intestines, swamps, marshes) grows along with the cattle population as well as with the area occupied by flooded rice paddies. At many oil wells, methane escapes without being captured or burned, and methane release (coaldamp) is an explosive hazard in some coal mines. Although only contraband production of CFCs continues today, the CFCs released between 1938 and 1992, or since then as refrigerators and air conditioners go to the junk pile, remain in the atmosphere and will continue to rise to the stratosphere for several decades yet. Difficult as it may be to foresee the future, it's not sticking my neck out to write that the greenhouse effect will continue to intensify during much of the 21st century.

GLOBAL WARMING?

The signing of the Kyoto Protocol at the end of 1997 signaled that decision-makers of many countries considered the risks associated with uncontrolled pursuit of greenhouse gas emissions to be unacceptable. On first examination, one might wonder what the fuss is about: even a doubling of CO_2

would trap only an additional 4 watts per square meter of infrared radiation in the troposphere (the lowest dozen kilometers of the atmosphere). If only this trapping is considered, the effect on temperature is small, less than 1°C (1.8°F) on average over the globe. But everything is connected in the atmosphere, and every perturbation perturbs everything. If this weak additional infrared trapping warms up the atmosphere a bit, and it must, the air can contain more water vapor, and any additional water vapor will further intensify the greenhouse effect. Does the real atmosphere work that way? In either hemisphere, water vapor content is indeed observed to increase going from winter to summer. Most specialists, but not all, consider that indeed the water cycle reacts to warming with such a positive water vapor feedback, amplifying by a factor 2 or more the response of the climate to carbon dioxide enrichment of the atmosphere. Furthermore, having perturbed the water cycle by carbon dioxide–induced warming, humans have at the same time perturbed the processes that form and dissipate clouds. That further complicates things: during daytime, clouds block part of the incoming solar radiation, but they block some of the outgoing infrared radiation both night and day. Moreover, the effect on the escape of infrared radiation depends on the height of the clouds. For some model calculations, the changes in cloud cover limit the warming caused by the CO_2 increase, but for others, on the contrary, they further amplify it. The uncertainty arises because such models cannot represent all the details of clouds or of evaporation from land surfaces, and different modeling groups use different approximations, no one being sure which one, if any, is right.

What has been observed over the last fifty or so years, as CO_2 concentration has risen from 310 to 372 ppm? It must be admitted that the climate changes have not been spectacular to most lay observers. However, they have been perceptible, and that in itself is remarkable because *climate*, as distinguished from *weather*, usually changes only very slowly. Fairly strong warming occurred between 1910 and 1950, followed by a quarter century of stagnation of the global average temperature; however, warming accelerated starting about 1975, and record temperatures were reached in the last years of the twentieth century. Although records don't go back very far, indirect evidence suggests that the world hasn't been so warm for at least one thousand years. Many specialists believe that the stagnation from 1950 to 1975, and the relative weakness of the warming observed since then, result from an effect that has in part compensated the intensification of the greenhouse effect. Burning of coal and oil containing small amounts of sulfur produces sulfur dioxide (SO_2), an unpleasant and unhealthy gas, which tends to form droplets of sulfuric acid in the atmosphere. The droplets do not accumulate indefinitely, instead they end up being scavenged by precipitation and falling as acid rain; but as more and more SO_2 is emitted, more and more droplets are present in the atmosphere at any one time. While in the atmosphere, they add to the reflection of solar radiation to space, and at the same time they may increase the reflectivity of certain types of clouds. This reinforcement of reflection of sunlight, reducing the flux of sunlight to the Earth's surface, what I like to call the "parasol effect," takes place for the most part downwind of major centers of industrial activity, wherever coal and oil are burned without worrying about air quality. High smokestacks protect nearby neighbors from the worst of the fallout, but if anything they make the large-scale problem worse because the pollution released higher stays longer in the atmosphere, albeit in less concentrated form. Since the 1980s, such pollution is no longer considered acceptable in North America and Europe. To comply with legislation (in the United States, the 1970 Clean Air Act, amended in 1990), "scrubbers" are installed to remove offending particles and gases (especially sulfur dioxide but also nitrogen oxides) from products of combustion or other industrial processes before allowing them to emerge from the smokestack. So far, however, such concerns have been considered low-priority in the developing countries. Also, until the 1990s and the collapse of "scientific socialism" in the Soviet bloc, sulfur dioxide pollution reigned over central and eastern Europe and further east in the USSR. With newly growing economies replaying older ones over the past half century or so, has such sulfur pollution partly masked intensification of the greenhouse effect? If indeed so, there is a risk that strong warming will emerge rapidly over the next decades, once the citizens of rapidly developing countries (such as China) decide to stop accepting to breathe poisonously polluted air. And that could happen well before their per capita fossil fuel consumption reaches American or German (or erstwhile Soviet) levels. In that case, the world's mean temperature could rise by as much as 4 to 6°C (7.2 to 10.8°F), an increase that in fact would be absolutely enormous, comparable to the transformation of climate and landscape that accompanied the end of the last ice age. But where that natural transition took thousands of years, strong greenhouse warming could come into play within several decades, with catastrophic consequences for the biosphere. Forests can and have moved in the past, but not in one century. Still, the worst is not certain, and with luck the scale and the speed of climate change may be manageable. How much should we trust to luck?

Specialists continue to argue about the extent to which intensification of the greenhouse effect has been masked by the reinforced parasol effect, and political motivations often infect the scientific debate. Some would argue that recent climate change demonstrates that most models quite simply overestimate the sensitivity of climate to greenhouse gas additions. The argument must be examined because, so long as its concentration does not exceed a few thousand ppm (and it is about 372 in 2001, although it often exceeds 1,000 in airliner cabins), carbon dioxide is no threat to health, and in fact it tends to enhance plant growth. However, the balance of evidence indicates that climate does indeed respond, on the global scale, to relatively weak perturbations. For the most part, over past millennia, the ice caps appear to have advanced and receded simultaneously in the two hemispheres, although the astronomical perturbation acted in favor of ice growth in one hemisphere only. That can be understood if atmospheric water vapor feedback reinforces the perturbation and transmits it to the other hemisphere. But if such feedback does indeed exist, it also must operate to amplify the effects of anthropogenic reinforcement of the greenhouse effect. The sources of water vapor remain unevenly distributed, as do the emissions of carbon dioxide, but the atmosphere respects no borders, and wherever it is emitted, CO_2 accumulates in the atmosphere all around the Earth. By the greenhouse effect of water vapor and carbon dioxide, the atmosphere transmits to all parts of the planet the response of climate to anthropogenic perturbations.

THE REAL RISKS

No, doubling CO_2 won't stop the world from going round. It could slow it down a bit, to the extent that water piled up in the form of ice caps near the poles may melt and end up in the oceans, closer to the equator. Such redistribution of the mass of water would increase the Earth's moment of inertia and slow its rotation, just as a figure skater can slow her spin by extending her arms (most of the time, she does the opposite). However, warming of the polar zones will not automatically melt the ice caps, because their volume or mass depends on the balance between snow being deposited on the ice cap, against ice flowing and melting at the ice cap's edges. The atmosphere above the poles contains very little moisture, but if its moisture together with atmospheric transport of moisture toward the poles were to increase, the ice caps could grow, which would lead to a fall of sea level, something that would create plenty of problems for seaports and shipping. Research continues on the subject, and recent results of calculations suggest that the Greenland ice cap would diminish as temperatures rose, but the Antarctic ice cap would change very little. Monitoring from space of changes in the ice caps only began recently, but it helps to fill in the picture. From time to time, a spectacular breakup occurs, and ice islands as large as several thousand square miles can drift away from the ice bank at the edge of Antarctica. Some specialists have worried that a large part of the West Antarctic ice sheet, weakened and destabilized by melting at its edge, could suddenly slide into the sea, provoking a catastrophic rise in sea level around the world by 4 or 5 meters in less than a hundred years. This may have happened many thousands of years ago, and it could happen again, but most specialists do not believe that it represents a significant risk for the next few hundred years. It is, however, true that the ice sheet has become a few hundred meters thinner in the last 10,000 years, contributing a little to the rise in sea level.

The problem remains, how should we deal with such a risk? What do we do about extremely unlikely eventualities that would however be catastrophic should they come about? For the effect of global warming on sea level, models predict a sea-level rise in the range from 30 to 90 cm (1 to 3 ft.) by the year 2100, mostly resulting from expansion of the warmer seawater rather than from melting of land ice. Understandably, the Alliance of Small Island States (AOSIS) considers even such a modest rise as totally unacceptable, as it threatens the very existence of its members. In Bangladesh, such a rise in sea level would enormously increase the cost of building the dikes needed to protect the multimillion population in the Ganges-Brahmaputra delta, dikes that do not exist today although they are needed whenever a tropical cyclone strikes. Along many coastlines all over the world, even a small rise in sea level would flood wetlands, which would be difficult to restore at the expense of owners of nearby still dry land.

Conjured up by sensationalist media reports, the specter of the sea washing over extended coastal plains scares some of the public, but that danger is not the most serious danger connected with global warming, at least for the next hundred years. It remains a real risk for future generations if greenhouse gases continue to be added to the atmosphere, if the world continues to warm, and if this leads to enough ice melt for sea level to rise by several meters. Health risks of global warming have also in my view been misunderstood. Most of the poor countries in the world today are situated in the Tropics, in the zone associated with heat, and they have far too many health problems. Still, why should we scare ourselves with the specter of tropical diseases arriving in the temperate zones along with warming? True, a few cases of malaria turn up around major airports during heat waves, and if the local physicians don't recognize them, it has serious consequences. But let's not exaggerate! Such cases, as also the recent arrival of the West Nile virus in the eastern United States, result from modern air transportation, and if they occur too often, the airlines will have to make sure that the mosquitoes don't make or don't survive the trip! It would be better not to have wetlands near major airports, and I would like to believe that no one will try deliberately to reintroduce the *Anopheles* mosquito in the name of restoring biodiversity. The world has warmed a fair amount since 1850, but malaria has been largely eliminated in North America and Europe. In the cooler nineteenth century, malaria was one of the risks of a visit to Rome, and it was common in many other regions of Europe, not only in the warmer south. Malaria remains one of the top killers in the world, affecting populations in poor countries, which also happen to be warm.

In the humid Tropics, and also in semiarid and arid zones, much of the population lives at the limit of survival, suffering from malnutrition and highly vulnerable to disease. Often, with very little freshwater available per capita, people lack any access to clean water. But thirst must be quenched, with contaminated water if no clean water is to be had, and this accounts for the high incidence of water-related disease. Rapid population growth necessarily increases the pressures on public health services as on all public services. As the state withers away, disease increases, as observed in Russia. Warming will no doubt bring new problems, and if the cooler countries do not keep up their guard, some mosquitoes and microbes, up to now blocked by nighttime or winter cold, may indeed appear. People will suffer, and some die, from more frequent and more intense heat waves, fewer will freeze to death. Which excess kills more in Europe or the United States? The real problems lie elsewhere.

A New Deal in Water

Intensification of the greenhouse effect will lead to significant global warming only if infrared trapping by added carbon dioxide is reinforced by positive water vapor feedback, with increased humidity trapping still more infrared radiation. That means that the principal danger of continued carbon dioxide emissions comes by way of changes in the water cycle. To talk only about global warming and sea-level rise is to ignore the complexity and gravity of the changes that may be in store. Water falls from the skies. This vital resource falls unevenly over the globe, depending as it does on the distribution of sunshine, the rotation of the Earth, the arrangement of the oceans, continents, and mountains over the globe, and the physical properties of water. History and its accidents have led to a situation where humans already use a large fraction of the water resource available in some parts of the world, much less in others. Enormously difficult problems face us in management of the resource so as to ensure enough clean fresh water for populations that will continue to grow for decades, even if the demographic transition goes to completion. Too often, thinking about water problems and planning of water infrastructures assumes implicitly or explicitly that water, the gift of the heavens, is a known constant element of the problem. The assumption is wrong.

On average, water resides only eight to ten days in the atmosphere, in the form of water vapor and clouds. Thermal inertia of the enormous mass of water in the oceans may delay climate change, but that delay affects only the evaporation of water, not its condensation. Up to now, only relatively weak warming has occurred, but if major warming takes place, it will be because humans will have perturbed the water cycle by way of CO_2 and in particular because the conditions of evaporation and condensation of water will have changed (fig. 6.1). And there lies the real danger, much more than in warming per se. Most specialists believe that reinforcement of the greenhouse effect and global warming will be accompanied by acceleration of the hydrological cycle, with both evaporation and precipitation being enhanced. Considering the way the atmospheric part of the water cycle operates, rainfall will increase in some areas, diminish in others. According to some model simulations, Hadley cell circulation will be enhanced, with still higher rainfall in the already wet intertropical convergence zone, and even more severe aridity around the tropics of Cancer and Capricorn. Will the dryness typical of North Africa extend to southern Europe? Water shortages already often affect Andalusia (southern Spain), and the prospect of more extended dryness is worrying.

In China, India, Pakistan, and in fact all around the world in the zone of transition between the humid and the semiarid Tropics, teeming populations eke a bare living from limited water resources. In other parts of the world—for example, on the Great Plains of America—large relatively dry areas, in part dependent on irrigation, produce the wheat that feeds the large nonfarming majority of the population. Around the Mediterranean, fruit production depends on winter rains and summer sun. Modification of the water cycle may lead to shifts in the well-watered areas and in the dryer border zones. Will the African Sahel, between the desert and the rainier zone to the south, come out winner or loser in the climate change lottery? In some regions, the changes may increase the water resource, easing the difficulties of supplying the population. But in others, the opposite may be the rule.[4] In many Third World countries, where supplying clean (or even not so clean) water to growing mega-urban populations is getting harder and harder, an unfavorable change in the water resource will change the crisis to disaster. That is the real threat to health.

More than global warming, the redistribution of the water resource brings with it great dangers in the coming century. Change need not be for the worse everywhere. In some areas, where rainfall increases, climate change may bring new hope. Also, with milder temperatures, frosts will be less frequent, and the length of the growing season will increase. Strangely enough, however, as discussed in chapter 8, the redistribution of water associated with global warming could possibly bring with it a crisis of cold in Europe, if additional freshwater were to block the great ocean heat-conveyor belt. Uncertainties abound, but can uncertainty be taken as an excuse not to act? Is it wise to accelerate the transformation of the atmosphere's composition, enhancing its greenhouse effect, knowing as little as we do about the consequences for water? Shouldn't we at least try to learn more? Sometimes scientists stand accused of crying wolf in order to get more funds for their research. That is no doubt part of the nature of things, but can one really make a case for ignorance?

Water falls from the skies, from the atmosphere, ever renewed and respecting no borders. For water, we worry first about our well, our river basin, and our watershed, but whenever we drink water or wine, whenever we eat bread or rice, we are using water that has come as rain from somewhere else on our wide world. In this sense, water is indeed a common human heritage, for which we must care with planetwide solidarity.

BACK TO THE ICE AGE

THE END OF THE PLEISTOCENE?

Can the present interglacial time be legitimately distinguished from preceding interglacial intervals of the Pleistocene? Geologists often call this the *Holocene*, the "completely new" epoch. It was in 1839 that English geologist Charles Lyell coined this term and others from the Greek to designate different ages of the Quaternary period: Pleistocene—the most recent, Pliocene—more recent, Miocene—less recent, and Eocene—the dawn of the recent. Lyell's completely new epoch now is part of our recent past, and in fact, the major distinction between the present Holocene and previous interglacial intervals—the most recent of which was 120,000 years ago—lies in the growing global impact of human activities. Without this new factor, the ice would almost certainly advance again, sometime in the next 20,000 years. Following that, for more than 100,000 years, climate could be expected to alternate between cold and very cold. After a new glacial paroxysm similar to the last one 18,000 years ago, ice would finally retreat and climate switch back to an interglacial interval similar to that of the last 10,000 years. Will the ice come back? Human activities *are* something new under the Sun. If indeed anthropogenic reinforcement of the greenhouse effect prevents the return of the ice, we shall indeed be entering truly new times, a Holocene epoch in the full sense, an epoch that should perhaps even be called "anthropocene."

It's hard enough to see clearly what is likely to happen in the next few decades; does it make any sense to theorize about coming millennia? If the ice caps return to Canada and Europe in twenty thousand years, there will have been ample time for humans to have exhausted all fossil fuels and to see (if any humans are still around to see) the CO_2 added to the atmosphere taken up again by vegetation and the oceans. With global warming induced by the anthropogenic intensification of the greenhouse effect, the ice caps may well melt completely away (but probably over a period of 3,000 years rather than 60 or 300) and sea level may rise, but since in past ages the ice returned many times to cover land from which it had retreated, why should it not do so again, once or twice or even twenty times?

From the Twentieth to the 21st Century

Enough time spent on the millennia. What about the next century or two? Thirty years ago, rather than greenhouse warming, some scientists feared a rapid return of the ice, with a sort of "snowblitz" covering large areas of Eurasia and North America with a strongly reflecting layer of snow too thick to melt away in the spring.[1] Since 1957, however, the measurements have revealed the continuing increase of atmospheric carbon dioxide, while over the same period temperature fluctuations switched from irregular to a marked warming trend since 1975. Since the 1980s, discussion of the risks associated with continued dependence on fossil fuels refers to scenarios and climate model simulations for the entire 21st century and in some cases up to the year 2400. I often wonder what people will think of our predictions then. Still, land use planning and investments in equipment and facilities having long lifetimes—dams, roads, bridges, power plants—must make a gamble on what the future will be like. Although predictions of change for the coming century are inevitably full of uncertainties, they are certainly not less likely than the implicit scenario of *absence* of change that the merchants of coal and oil would have us take for granted.

Climate models cannot yet give reliable predictions of how distributions of rainfall, evaporation, and soil moisture will change, even for a date as close as 2050. To reduce the uncertainties calls for sustained monitoring and research, especially since improved understanding of what becomes of the water falling on the land will make it possible to use the water resource more effectively. In fact, enormous advances *have* been made. In 1900 wheat production barely exceeded one ton per hectare (about two tons per acre). And in India, where wheat production was only 0.83 ton per hectare in 1965, the "green revolution" multiplied total production by a factor 5 without doubling cultivated land area. Even today, however, less than one ton per hectare is produced in many dry areas of Africa. Chemical fertilizers count as a major contribution of modern technology to such agricultural productivity gains not requiring additional water. Of course, that involves new risks, especially when too much fertilizer is applied, and it does not eliminate the need for water; optimum application requires tradeoffs between protection of the water table, increase of agricultural productivity, and safeguard of tomorrow's soils.

Let's not forget another contribution of technology to production of food for the human population. Tractors have taken the place of horses, mules, donkeys, and oxen, who used to do most of the heavy work, and large draft animal populations are no longer needed. In 1910 production of animal feed required 30 percent of the farm acreage in the United States. Of course, populations of cattle, sheep, and pigs continue to increase to satisfy growing demand for meat and milk, but farmers in advanced countries no longer need to set aside a significant part of their production to feed draft animals. In the Third World countries, by contrast, hundreds of millions of oxen, water buffalo, camels, donkeys, horses, and mules eat food that could otherwise feed 700 million persons. Note also that production of meat eaten by a fairly small minority of the world's people requires food that could otherwise feed some three billion people. To each of these figures corresponds a figure giving the needs in water and farmland. Only by increasing the amount of food produced per hectare (or acre, or *dunam*, or *feddan*) has it been possible to nourish (not always well) the growing human population, without farming the entire land surface of the planet. Have these enormous gains in productivity been attained without damage to the "sustainability" of agriculture, by erosion or by deleterious chemical alteration of the soil? Some agronomists believe that productivity can still be multiplied by another factor 5, whereas others believe that modern agricultural methods cannot safely be pursued for many decades more. A vital debate.

Another problem for the future: modern technology depends on harnessing energy, to a large extent by exploiting *non*renewable resources;

moreover, it involves new insults to the human or natural environment. We no longer have to clean up tons of horse manure from the streets of our cities and towns, and the clatter of hooves on the pavement has gone, but obstinate insistence on using private automobiles, even in crowded cities, often brings as much or more noise and stink. Outside cities, more and more broad highways crisscross fields and forests and make it harder and harder for whatever wild animals are left to move around. Instead of growing fodder for horses, we import oil. Even those who don't take seriously the risks of intensification of the greenhouse effect must recognize that our dependence on black gold has its own costs, both for the environment (consider the oil spills by *Exxon Valdez* off Alaska, by *Amoco Cadiz* off Brittany) and for society as a whole (the Gulf War). It will take millions of years to renew the oil resource we dilapidate so rapidly. Much more coal still lies in the ground, but mining scars the land and takes lives, and exploitation of some resources such as the Colorado oil shale is impractical considering limited available water resources, at least with today's technology. Can the world's energy needs be satisfied without expanding the use of nuclear power? Can enough energy be produced from renewable resources, essentially solar, including biomass burning (but water is needed to grow the plants to be burned) and hydroelectric power from the many still unexploited high-potential sites (as well as wind and waves)? Thermonuclear fusion produces the Sun's energy, but on Earth is controlled fusion a real option, or a mirage? In any event, it should not be forgotten that for the next few decades, the energy source most easily harnessed, at least in advanced countries, is today's wasted energy. With truly modern technology, with enough political will to do away with hidden subsidies of outmoded industries and fossil fuel producers, we can do more with less, perhaps ten times less.[2]

FINITE AND INFINITE RESOURCES

World population will almost certainly pass the eight billion mark sometime in the 21st century, and it may approach twelve billion. Such large populations can be fed, but in the absence of additional water resources, without new ways to multiply water use efficiency, agriculture will have to take over practically all the world's land and tap all the world's flowing freshwater. How can anyone find such a prospect acceptable? Even apart from nature lovers nostalgic for wilderness unsullied by humans, how many people really want to live in a world that would end up being nothing but a gigantic agro-industrial complex? Increasing the water resource is conceivable, in particular by going after water flowing in rivers considered inaccessible up to now. The rivers flowing north to the Arctic may well be exploited before the end of the 21st century, although plenty of questions remain unanswered concerning the consequences of such freshwater withdrawals for Arctic ice cover and snowfall on surrounding lands. Hardly anyone evokes anymore the various grandiose projects for the Congo River and Lake Chad basins, and the idea of flooding the vast Saharan Desert areas below sea level. In 1910 such schemes were promoted with the idea of "enhancing the value of Africa as a land for settlement and colonization by Europeans," without asking what the Africans thought of the idea. In another context, even as late as 1977 there were optimists to write: "During the years of Soviet power, many areas that were 'dead' or unproductive deserts only a few decades ago have been turned into flourishing districts of developed industry and agriculture."[3] Today, such problems are approached in a new light, both from the political and from the technical and scientific points of view, and specialists and lay citizens alike are more aware of the snares and traps of taming nature. Still, we can learn a lot over the coming century, and the dream of making deserts bloom may well become a more common reality. Not that I would approve of greening all deserts. The stark bright desert landscape, a precious part of our heritage as inhabitants of this planet, reveals clearly the past history of running water on the land. But deserts cover vast territories, and gardens, like the oases created as stopping places for caravans carrying salt in exchange for gold across the Sahara, have their own, human-created beauty. Eden was a garden, not a forest—a God-created garden (Genesis 2:8), but at the same time a garden created by Man (and Woman), to whom the Creator gave the task of working the land so that it would bear fruit.[4]

How can we make better use of our resources? The Sun is always shining on the Earth, and the Earth must get rid of the solar energy that it absorbs and turns into heat by radiating it away as infrared to space. However, our planet's water cycle operates in a practically closed circuit, transforming and transporting an enormous (but by no means infinite) stock of water. A space station also operates in this way, with a much more limited water stock; and as cities grow, they will end up by having to operate that way too, as already should factories. Up to now, agriculture has been very different, because even though water transpired by crops ends up in the planetary cycle, it returns to the land only after the long detours of atmospheric circulation, cloud formation, and precipitation in convective or cyclonic disturbances. Certainly the water resource can and must be used more efficiently—for example, using drop-by-drop irrigation to keep evaporative losses to a minimum and to make sure the water benefits the crop. But since we are worried about the greenhouse effect, why not one day grow crops with a much shorter water cycle entirely confined within a greenhouse? Even if such greenhouses were to be horribly technological facilities, wouldn't they make it possible to leave larger areas of forest, prairie, or desert in their "natural" state? Today, the idea of feeding billions of humans with crops thus grown may appear to be a pipe dream or science fiction, and it may be another crazy idea of a theoretician who has never built a greenhouse, but tomorrow, if we can develop sufficiently light-weight and sturdy materials, transparent to sunlight but opaque to infrared rays, why not? For the most part, twentieth-century science and technology went far beyond the "reasonable" predictions made in 1900, and compared with what is happening with computers and biotechnology, thousand-square-mile greenhouses may one day be considered almost ordinary!

AVOIDABLE CATASTROPHE

Doomsayers have always been with us. In 1948, when Americans were preparing CARE packages (Cooperative for American Remittances to Europe) to help starving European populations to pick up after the disasters of World War II, New York ornithologist William Vogt published a best-selling book, *Road to Survival*, in which he argued that Europe as well as Asia would have to reduce their populations substantially if they were to survive. And time and time again, distinguished voices have proclaimed that we are on the way to famine, that the "death of the oceans" is nigh. But catastrophe is not inevitable, except for those who consider our very existence as human beings on this planet to be a catastrophe. I can empathize with nostalgia for an imagined lost state of nature, but I know that nature unsullied by humanity disappeared more than 8,000 years ago when farming began. I also am convinced that the end of that nature has created possibilities that we humans would not gladly give up. There's no way back. We humans are condemned to exercise stewardship over the Earth, managing the land and the water, and we can do it and protect nature at the same time. But we must accept that a protected nature, even if not a human creation, requires at least that we humans work together with nature and accept our responsibility for it.[5]

CHAPTER 15

CONCLUSION—THE END OF THE STORY?

BACK TO THE BEGINNING?

Since at least three billion years, liquid water has existed on our planet, making possible the evolution of all forms of life. The Sun has grown brighter over the same period, but the oceans have never boiled. Surrounding our never completely frozen Earth, the atmosphere's powerful greenhouse kept the oceans liquid when the Sun was weak. As the Sun grew brighter, gradual weakening of the Earth's atmospheric greenhouse ensured the stability of climate and favorable conditions for the development of life. Ice sometimes covered enormous areas, but the ice always receded, sometimes disappearing almost completely from the planet's surface. Sea level fell and rose again, and at least once in their long history the Mediterranean and the Black Seas evaporated completely, leaving only deserts of salt behind; but the water always returned.

One day this long and wonderful story will end. The slow steady increase of the Sun's brightness results from changes in its internal structure, changes made necessary by the inexorable conversion of hydrogen to helium at the center of our star. As hydrogen available for "burning" runs out, energy production in the central core decreases, temperatures and densities remaining too low for burning of helium to form carbon. The Sun keeps shining, its energy production now taking place mostly in an intermediate envelope outside the central core, and as this process accelerates over the next four billion years, the Sun's outer envelope will have to expand; our star will become brighter and brighter. As it uses up its last hydrogen fuel reserves, it burns faster and faster, brighter and brighter. How can the Earth's atmosphere and biosphere adapt to the Sun's evolution? Over the last few decades, human activities have been reinforcing the greenhouse effect and warming slightly the surface and lower layers of the atmosphere of the Earth. But this process, although apparently too slow for many politicians to apprehend, will be but a brief parenthesis in the long history of Sun and Earth. Over the last few billion years, the planet has kept cool by weakening the greenhouse effect as the Sun grew brighter; over the next few billion years, the problem for the planet will grow more acute as the Sun grows still brighter. Can the greenhouse effect be weakened still further? And for how long? According to the "Gaïa" hypothesis of English chemist James Lovelock and Boston University biologist Lynn Margulis, life maintains optimal conditions for its existence and development by modifying the physical and chemical properties of the Earth and, in particular, the composition of the atmosphere. Indeed, the biological process of photosynthesis gradually transformed much of the CO_2 initially present in the atmosphere into organic matter, and the "sequestration" of much of this matter as sediments and as fossil fuel underground explains the long-term reduction of atmospheric CO_2. The resulting weakening of the greenhouse effect seems to have compensated the increase of the Sun's brightness. For how long can such a process work? Human activities have been putting some of the fossil carbon back into the atmosphere as CO_2, but even if the level is quadrupled for a few centuries, that will not make much CO_2 to be extracted from the atmosphere. Moreover, photosynthesis in its present form could not continue if all the CO_2 were to be removed from the atmosphere. In any event, carbon dioxide is only the second greenhouse gas, the most important being of course water vapor. Climate sensitivity to perturbations depends on many feedback mechanisms, some of which act to amplify change. Positive water vapor feedback plays a central role in determining how much global warming will result from anthropogenic emissions of greenhouse gases, in particular CO_2, over the next few centuries. However, positive feedback will also amplify global *cooling* in response to a *decrease* in CO_2. With luck, a beneficent Providence, or geo-engineering by earthlings far more advanced than today's humans, the Earth system may be able to maintain photosynthesis while at the same time avoiding excessive heating by a brighter Sun, by some appropriate combination of changes in the factors of CO_2, atmospheric humidity, cloud cover, and surface snow and ice cover. For Earth to remain livable a billion or two years from now, clouds or ice must ultimately win out over the greenhouse effect.

Even if life's lease on Earth can be extended, it will come to an end when the Sun becomes a red giant, its outer envelope approaching the orbit of our planet. When that happens, temperatures will have to rise to the point where the oceans begin to boil. Earth's waters will evaporate and escape to space. Will those H_2O molecules meet again, some day on some other planet around some other sun? Will the story start all over again? We can always dream . . .

Have a drink of water. If you fill your glass with the excellent quality tap water of Paris, you'll swallow some drops that fell as rain near the sources of the Seine, perhaps as recently as last winter, perhaps several years or even a decade or two ago.[1] That precipitation came from clouds most probably formed over the Atlantic from water vapor evaporated from the Gulf Stream or from other not-quite-so-warm water drifting northward. At least some of that water has a history of centuries of circulation in the depths of the ocean, between the Norwegian Sea and the Indian Ocean or even the Pacific. If, however, you'd rather quench your thirst with mineral water, the recent history of your drink, before it was bottled, may go back thousands of years, with a long detour in glaciers and underground. In both cases, many of the water molecules have been exchanged billions of times between the oceanic Earth and the skies, ever since the primordial outgassing of the planet's innards. When you have a drink of water, you are absorbing, for a brief time, a small sample of matter carrying with it part of the history of the Earth and of the universe.

You'd rather have some wine? Fine! To the water evaporated from the Atlantic that fell as rain on the vineyards around Bordeaux, nature and the winegrower have added sunshine, CO_2, loving work, and a dash of other elements, stardust. *A votre santé!*

NOTES

PREFACE

1. Meteosat-1, the first European geostationary weather satellite, was launched in November 1977 from Kourou, near the equator in French Guiana. This came a few years after the first American geostationary, but it was the first to include a water vapor channel, i.e., to observe at infrared wavelengths at which the water vapor in the atmosphere prevents radiation from the surface from escaping to space. Like all geostationary satellites, whether for Earth observation or communications relay, Meteosat is in a circular orbit at an altitude of 36,000 km (22,400 mi.), moving around Earth in the same time that it takes our planet to make a complete rotation. Its normal position is above the intersection of the equator and the Greenwich meridian (longitude 0). The Meteosat series has been extremely successful, and spare models kept in reserve have at times been shifted to other longitudes.

2. In the original: "Mit Eifer hab ich mich der Studien beflissen; zwar weiss ich viel, doch möcht ich alles wissen." I remember this line so well because my mother would often quote it and other lines from Goethe, which she remembered from her school days in Berlin. But sometimes I wonder if these particular lines were not to make fun of me.

PROLOGUE

1. The same is true with the Greek *pneumatikos* for the Holy Spirit as well as ordinary breath, and in words derived from Latin such as *spirit* and *respiration.*

2. If it appears politically incorrect to write of man and his mastery, let me object that men have played at being masters long enough to take the blame for their mistakes. But women have nearly always been the ones to carry the water from the wells.

3. This is the explanation often given for the decline of the Harappan civilization in the Indus valley, or that of the Sumerians along the Euphrates (Jacobsen 1958), but it is not the only one. Other environmental disasters have been invoked, such as catastrophic floods subsequent to a major earthquake.

4. Now that we see satellite images of Earth on television weather channels every day, we've all become extraterrestrial observers. Of course, Venus is entirely cloud-covered, as are the giant planets (Jupiter, Saturn, etc.); clouds are rare on Mars, and totally absent on Mercury. On Earth, clouds are common but not constant; the extraterrestrial (or satellite) observer who wants to see the surface (for commercial, military, or other reasons) has good chances of doing so, but must take clouds into account.

5. As noted by Alistair C. Crombie (1959:129), the idea that rain is the source of the water for springs and rivers was developed earlier by Conrad von Megenburg (1309–1374) and goes back to the Roman architect Vitruvius in the first century B.C.

6. Bernard Palissy (1510–1590) had many artistic accomplishments (architecture and ceramics, in particular polychrome enamelware) in addition to his writings on "natural philosophy." Although he lived to eighty, he died in terrible circumstances in the Bastille jail after a last-minute reprieve from execution as a "heretic." It was risky to be a Protestant and a free spirit in sixteenth-century France. A street in Paris is named after him.

7. Louis Joseph Gay-Lussac (1778–1850) also has his street in Paris. So do Antoine Lavoisier (1743–1794) and Henry Cavendish (1731–1810) as well as Berthollet, Fourcroy, and Guyton de Morveau.

8. At standard conditions of atmospheric temperature 14°C and pressure 1,000 hectopascals (or millibars).

9. After the definition of the metric system, a standard bar of length one meter was fabricated (and it is still kept at the International Bureau of Weights and Measures, in Saint-Cloud, just outside Paris). Later, it was decided to maintain that bar as defining the meter even though new measurements slightly changed the size of Earth expressed in those meters. Also, the planet Earth is not a perfect sphere. The circumference around the equator is 40,077 km; along a meridian, passing through the poles, it is 40,032 km. Note that the changes amount to less than 0.2 percent of the 40,000 km (about 25,000 statute miles, where a mile was originally the Roman unit of 1,000 paces). The Earth's circumference around the equator is 21,600 nautical miles (360×60, since 1 nautical mile is by definition exactly one arc minute). However, the British Admiralty and the U.S. Coastal Survey disagree slightly on the exact value in feet.

10. These issues are explored at a nonspecialist level in the book by Philip Ball (1999), and new research results continue to appear.

11. Mediterranean water also exits that way to the northeast. In their book *Noah's Flood*, William Ryan and Walter Pitman (2000) note that sailors heading north in the Bosphorus (between Istanbul and Asia Minor) once used weights suspended from their boats into the current of salty Mediterranean water flowing along the bottom of the strait toward the Black Sea, to help them fight the surface current flowing the other way.

1. BEGINNINGS

1. Fred Hoyle (1920–2001), English astrophysicist, cosmologist, and science fiction writer, never was convinced of the Big Bang (Hoyle, Burbidge, and Narlikar 2000).

2. The development of modern cosmology is narrated in many popular books and college texts on astronomy and the history of science. See, for example, Munitz (1957), Weinberg (1993).

3. American architect Richard Buckminster Fuller (1895–1983) pioneered the use of light and easy-to-assemble structures for enclosing small and large spaces. He formulated the geodesic dome concept in 1942, based on abstractions of pure geometry. The discovery that nature used the same design on much smaller scales came much later: an example of what Paul Davies (1993) calls "The Mind of God"?

2. THE CHURNING OF THE EARTH

1. David S. McKay and colleagues at the NASA Johnson Space Center found microscopic structures inside meteorite ALH48001, known to have reached the Earth coming from Mars, which they identified as fossil Martian bacteria. Other researchers cite contamination problems and certain details of the meteorite's mineralogy as evidence against this (Gibbs 1998).

2. The Mars science fiction trilogy of Kim Stanley Robinson (1993) treats this theme, developed as a possible real-life scenario by Christopher P. McKay (1999) of the NASA Ames Research Center.

3. Of course, there have been some disappointments (1999–2000), with the grotesque failure of one mission because of inconsistency between the metric units used by NASA scientists and the "English" units dear to industry engineers, and the still unsolved mystery of the last lost Mars Lander. But continued analyses of data from unmanned satellites put in orbit around Mars confirm that there were once oceans on Mars, and that in at least one area liquid water may have emerged at the surface quite recently, i.e., thousands or millions rather than billions of years ago (Zorpette 1999; Malin and Edgett 2000). Although more and more data will be sent by robot systems, the urge to send men and women to Mars may yet prove irresistible before the first half of the 21st century is over.

4. The NASA interplanetary probe *Galileo,* which first entered the Jupiter system in December 1995, provided key data on the oceans of Europa (Pappalardo, Head, and Greeley 1999).

5. Already in the seventeenth century, satirist and playwright Cyrano de Bergerac (1619–1655) wrote about life on other worlds; best known as the swordsman hero of Edmond Rostand's play of 1897 (also a movie), Cyrano really existed. In 1686, Bernard de Fontenelle (1657–1757) published a volume of *Conversations on the plurality of inhabited worlds* with Madame la Marquise de G****). See also Shklovskii and Sagan (1966:2–10).

6. In 1999, uncontaminated inclusions of extraterrestrial water in liquid form were identified in a meteorite fallen in Monahans, Texas, thanks to the quick reaction of the boys who found the meteorite and brought it without delay to the NASA center in Houston.

7. Predicting just how bright a comet will be is very risky, especially if the predictions are exaggerated in the media. There have been dismal disappointments and spectacular surprises. In the spring of 1997, I was astounded to be able to see comet Hale-Bopp in the sky amid the lights of downtown Paris from my apartment terrace.

8. An "excited" atom is one in which one or more of the electrons is found in a higher orbit than usual, the "kick" coming either from radiation or from a collision with another atom or other particle.

9. See, for example, Dessler (1991), Deming (1999), and Frank and Sigwarth (1999).

10. The question of a (nearly) completely ice-covered "snowball Earth" is again a "hot" topic, after discovery of geological evidence of low-latitude sea-level glaciation some 400 million years ago, that only a part of the then-tropical oceans remained ice-free, and that this situation played a major role in biological evolution before violent and prolonged volcanic activity put an end to it by copious emissions of greenhouse gases (Hoffman and Schrag 2000).

11. This theme was first developed in modern terms by Ichtiaque Rasool and Catherine De Bergh (1970), then at NASA's Goddard Institute for Space Studies in New York.

12. The French title—*L'Ecume de la Terre*—meaning "the scum of the Earth," was not retained in the English translation (Allègre 1988). Claude Allègre, who won the distinguished Crafoord Prize in geology in 1984, more recently made the news in his role as French Socialist minister of education and research, working hard to introduce some necessary reforms but ending up being thrown to the wolves.

13. Similarly, the death toll was very high in the Izmit (Turkey) earthquake of August 1999. Although the building codes took proper account of the seismic risk, they were not well observed in too many cases. India was badly hit in January 2001.

14. The death toll (several hundred) was not negligible even so, but without evacuation it would have been far worse. Of course, there was no way to avoid the effects on the environment and the atmosphere, and the slight blockage of sunlight by particles resulting from the eruption partially reversed the course of global warming over the following two years.

15. More recently, on July 17, 1998, a local tsunami drowned thousands in Papua New Guinea. Frank González (1999) discusses its causes.

3. ORIGIN AND EVOLUTION OF LIFE

1. Swedish chemist Svante Arrhenius (1859–1927), who made important contributions on electrolytic dissociation, was the first to raise the question of global warming by reinforcement of the greenhouse effect due to fossil fuel burning (at the time, in 1895, about 500 million tons of carbon per year, more than ten times higher today). The warming of a few degrees that he calculated for the doubling of CO_2 in the atmosphere is quite close to modern estimates.

2. Indeed, Thomas Gold (2000) has argued that such heat-loving bacteria populate a massive "deep hot biosphere."

3. American Harold C. Urey (1893–1981) won the Nobel Prize in Chemistry in 1934 for his discovery of the deuterium isotope of hydrogen, and of heavy water (HDO and D_2O). During World War II, his group at Columbia University developed the science of uranium isotope separation as part of the Manhattan Project. He made numerous important contributions to the science of planetary atmospheres, so that many astronomers consider him as much an astronomer as a chemist.

4. Antoni van Leeuwenhoek (1632–1723) first earned his living in Holland as a draper, but later turned to playing with lenses. Although he never knew much about natural history, his invention of the microscope and his precise descriptions of the marvelous world of the small earned him the title of "father of protozoology and bacteriology."

5. I here allude to the idea that a monkey typing at random long enough would end up producing all of Shakespeare's plays and indeed everything ever printed using that character set (but a real monkey would get bored long before, and even if not, the world might end).

6. In his book *The Double Helix*, American biochemist James D. Watson (1981; originally published 1968) describes vividly the race for discovery of DNA structure, for which he and his English colleagues Francis Crick and Maurice Wilkins, working at the Cavendish Laboratory (Cambridge, U.K.), received the 1962 Nobel Prize in Chemistry. Many observers believe that Rosalind Franklin, working in the same team, should also have received a share in the prize.

7. See Danchin (1990).

4. CATASTROPHES

1. Pierre Teilhard de Chardin (1881–1955), French Jesuit priest, geologist, paleontologist, and philosopher, made important discoveries on the origins of modern man during his long stay in China. He later popularized a personal philosophy of evolution. Although initially the Roman Catholic church viewed his

writings with disfavor and forbade him to teach in Catholic establishments, it eventually relented.

2. See Gould (1991).

3. For a philosopher's exploration of the role of chance in the world and the affairs of man, see the book *Luck* by Nicholas Rescher (1995).

4. Erwin (1993, 1996) has argued for lowered sea level as the cause of this extinction, but also considers extraordinary volcanic activity as a possible factor. Dramatic release of methane has also been discussed.

5. Genera: plural of genus. In biological classification, the *genus* is the division just below the *family* and just above *species*. Thus, we humans are in the Animal Kingdom, in the Phylum of *Chordata*, in the Class of *Mammalia*, the Order of *Primates*, the Family of *Homonidae*, Genus *Homo*, Species *Homo sapiens*.

6. See Shklovski and Sagan (1966:100), still a very enjoyable book.

7. Some specialists estimate the age of the boundary at 68 million years, some at 65 million. The exact value can be argued about, but what is important is the fact that whatever the event was, it changed things radically in a relatively short time, less than a million years.

8. One often speaks of volcanic dust, and indeed there can be a lot of it in the air immediately after an eruption, and on the ground not much later. What counts for climate is the veil of particles (aerosols) that form and remain suspended for long periods in the stratosphere, where the particles can't be rained out. The most important of these are sulfuric acid crystals and other sulfate particles formed from the sulfur dioxide gas (SO_2) exhaled in the eruption.

5. ICE, MOON, AND PLANETS

1. From the story by H. G. Wells (1927):

You see, when Mr Fotheringay had arrested the rotation of the solid globe, he had made no stipulation concerning the trifling movables on its surface. . . . So that the village, and Mr Maydig, and Mr Fotheringay and everybody and everything had been jerked violently forward at about nine miles per second—that is to say, much more violently than if they had been fired out of a cannon.

2. Although the Red Sea is usually given as the Exodus crossing site, most scholars note that the Hebrew *Yam Suf* is better translated as the *Sea of Reeds*, probably referring either to a shallow northward extension of the Red Sea or to one of the Bitter Lakes further north. It was much later that the battle of Jericho was fought and prolonged by Joshua 10:12: "Sun, stand thou still upon Gibeon; and thou, Moon, in the valley of Ajalon."

3. Some scientists were so incensed that a serious university press could publish such a book that they threatened a boycott of that press's textbooks, a type of

pressure that created quite a scandal at the time. Macmillan took the book and made a bundle on it.

4. The title of the French equivalent of the *Nautical Almanac*, prepared by the astronomers of the Observatory of Paris and the Bureau des Longitudes is the *Connaissance des Temps* ("knowledge of time").

5. For a given pair of masses, gravitational attraction varies as the inverse square of the distance (10 times closer, 100 times stronger); but the tidal force, a small difference in gravitational attractions, varies with the inverse cube of the distance (10 times closer, 1,000 times stronger).

6. The great English poet Alexander Pope wrote the following epitaph for Newton's tomb in Westminster Abbey:

Nature and Nature's laws lay hid in night.
God said "Let Newton be," and all was light.

7. They also provide information on the rise (or fall) of mean sea level relative to the land. In general, sea level has been rising, but in some places this is exaggerated by the sinking of the land, while in other places (such as Scandinavia) sea level appears to be falling because in fact the land has been rising by rebound of the crust since the melting of the heavy ice cap.

8. Mills harnessing tidal power on a small scale go back practically a thousand years, and some ambitious tidal power plants were proposed as early as 1935 (Gibrat 1966). However, the Rance facility is the only one in the world supplying significant power to an electric power grid, and even its output (less than 100 MW or megawatts on average) is quite small compared to a typical nuclear (or coal-fueled) power plant.

9. *Confessions of St. Augustine* 7.6, translated by F. J. Sheed (New York: Sheed and Ward, 1942), 112.

10. Pierre Louis Moreau de Maupertuis (1698–1759), a pioneer of analytical mechanics, led the Lapland expedition. The results refuted the arguments of Paris astronomer François Cassini (based on Descartes' mistaken ideas and a survey of the meridian from Dunkirk to Perpignan in France) that the Earth was elongated rather than flattened as it is.

11. Very freely translated by me from Voltaire's 4th discourse on Man ("De la modération"):

Héros de la Physique, argonautes nouveaux,
Qui franchissez les monts, qui traversez les eaux,
Ramenez des climats soumis aux trois couronnes,
Vos perches, vos secteurs, et surtout deux Lapones,
Vous avez confirmé dans ces lieux pleins d'ennui,
Ce que Newton connut sans sortir de chez lui
Vous avez arpenté quelque faible partie
Des flancs toujours glacés de la terre aplatie.

12. French philosopher Auguste Comte wrote (1835:32–33): "It is certain that changes in the distance of the Earth from the Sun, and consequently in the duration of the year, in the obliquity of the ecliptic, etc., which in astronomy would merely modify some coefficients, would largely affect or completely destroy our social development" (my translation).

13. Translated by me from Milankovitch (1920:v–vii).

14. The chaotic nature of the obliquity of Mars, ranging from practically zero to values as high as 85°, appears to be well established (Ward 1973; Laskar and Robutel 1993). The Orbital Camera on the *Mars Global Surveyor* orbiting the planet revealed details as small as several meters, providing much new information on ancient and recent volcanic activity, on groundwater formation, ancient floods, and ice formation (Malin and Carr 1999).

15. The stabilizing effect of the Moon has been shown by calculations made at the Paris Bureau des Longitudes (Laskar, Joutel, and Robutel 1993). This contradicts the hypothesis that the obliquity of the Earth's ecliptic was once much larger than now, proposed to explain how ice could have covered sea-level areas close to the equator some 2.2 billion years ago. With high ecliptic, seasonal changes would have been enormous, but on average, conditions would have been colder at the equator than at the poles, permitting ice at the equator but not over the whole Earth. With obliquity stabilized close to the present value, it is on the contrary difficult to avoid the conclusion that the Earth was then *completely* covered by snow and ice, and the problem then is to explain how life survived and how such a "snowball Earth" could warm up afterwards.

16. Compare, for example, Goudie (1977; esp. ch. 7), Mason (1976), and Imbrie and Imbrie (1979).

17. Jacques Labeyrie (1993) has emphasized the changes in the Nile valley and in the Persian Gulf. Very recently, oceanographers William Ryan and Walter Pitman (2000) of Columbia University's Lamont-Doherty Earth Observatory have argued that the story of the Flood corresponds to a much more dramatic and cataclysmic event, the sudden flooding some 7,600 years ago of the present-day Black Sea (then a freshwater lake below sea level) by a breakthrough of waters from the Mediterranean that had been gradually rising for a few thousand years along with sea-level in general.

18. These results were first obtained at the glaciology laboratory of Grenoble in France (see, for example, Petit et al. 1999 for the definitive report on the Vostok ice core) and confirmed by laboratories in Bern (Switzerland) and elsewhere. A major difficulty is making sure that the bubbles of ancient air are not contaminated by modern air between the ice core extraction and the laboratory analysis.

19. Rereading this as I translate, I find that I may have overemphasized the potential role of the atmosphere (but the book is about "water from heaven"!) and neglected the oceans, which also "[bathe] Greenland, Europe, North America, the Tropics, and Antarctica." For the most part, changes in the circulation of the oceans are slower than atmospheric changes, no doubt about it, but new results suggest

that some features of the deep ocean circulation can change very quickly indeed. I'll come back to that in part II.

6. WATER AND ENERGY CYCLES

1. Figures are based on my interpretation of the much more complete (but sometimes mutually inconsistent) tables to be found in Baumgartner and Reichel (1975), Berner and Berner (1987), Cohen (1995), Herschy and Fairbridge (1998), and UNESCO (1978).

2. This includes about 20 W/m^2 of infrared radiation emitted by the surface and transmitted to space by the atmosphere. Estimates made as early as 1900 turned out to be reasonably good, but measurements of this radiation flux to space only became possible with instruments on artificial satellites beginning in 1960.

7. WINDS, WAVES, AND CURRENTS

1. English scientist Edmund Halley (1656–1742, better known for "his" comet) published the first world map showing the trade winds in 1686, and sought to explain their link with solar heating. George Hadley (1685–1758), also English, improved on Halley's work, attempting to take account of the effect of the Earth's rotation. See Frisinger (1983) and Hamblyn (2001).

2. French mathematician Gustave Gaspard Coriolis (1792–1843) developed the rigorous theory of the effect of the Earth's rotation on atmospheric motions in 1835.

3. Alexander von Humboldt (1769–1859), who was as much at ease in Paris (where he was member of the Academy of Sciences) as in Berlin, spent several years of his life exploring Hispanic America, where numerous geographical features are named after him (Meyer-Abich 1969). His *Cosmos* laid the foundations of modern climatology and indeed of much of geophysics. The point about climates differing between the eastern and western shores of the Atlantic figures in Humboldt (1846:378–79).

4. In fact, Walker did not limit his investigations to the South Pacific. He also defined and named the North Atlantic Oscillation (about which more later) and a North Pacific Oscillation.

5. The U.S. National Weather Service as well as research groups at Colorado State University and at the International Research Institute for Climate Prediction (organized jointly by Columbia University's Lamont-Doherty Earth Observatory and the University of California / San Diego) now engage in El Niño–La Niña forecasting, with increasing success. However, after driving away El Niño in summer 1998, La Niña stayed longer than expected. As of September 2001, the prognosis was for El Niño to return at Christmas and stay on with moderate strength in 2002.

In fact, El Niño did not return for Christmas in 2001, but as of June 2002 he seems to be well on his way for December 2002. By the time this appears in print, we should know how vigorous he is.

6. Biologist and oceanographer Rita Rossi Colwell, director of the (U.S.) National Science Foundation, has presented evidence (1996) that cholera pathogens proliferate under El Niño conditions.

7. It should be noted that El Niño conditions do not rule out particularly damaging hurricanes. Although for the most part, mild El Niño conditions prevailed from 1990 to 1995, hurricane Andrew, passing close to the center of Miami in 1992, caused record damage in terms of dollars. Very recent analysis (Goldenberg et al. 2001) warns that the next few decades may see many more (and more active) hurricane seasons than were observed in the 1980s and early 1990s, with a potential for enormous increase in damages in the Atlantic-Caribbean sector.

8. As I was working on this English-language version in April 2001, the land was completely soaked in much of France, with flooding in some areas, as a result of record rainfall in the last quarter of 2000 and the first part of 2001—one rainstorm after another coming in from the Atlantic, while temperatures remained mild through nearly all the winter.

9. This is a modern version of Walker's North Pacific Oscillation (note 6 above), and recent work relates it and the NAO to an *Arctic Oscillation* affecting the entire circumarctic zone.

8. WATER'S DEEP MEMORIES

1. In 1988, medical researcher Jacques Benveniste working at Clamart near Paris claimed (in a paper published with reluctance by the prestigious scientific weekly *Nature*) that water did indeed retain some sort of memory that could account for claimed effects of extremely dilute solutions of other substances (solutions so dilute that no molecules of the other substance were left in the small samples considered). This briefly made the front pages of Paris papers, but was never confirmed or accepted by other scientists.

2. Some of the first results were obtained by a Danish-U.S. team at Camp Century (Greenland), and recent work on Greenland ice cores involves both American and European teams. However, the Soviet (later Russian) Antarctic research station Vostok has furnished the longest record so far (Petit et al. 1999), and indeed the ice core extraction was halted only because of reluctance to risk contamination of Lake Vostok, a remarkable body of liquid water deeper under the ice.

3. There's not much at Rockall apart from some rocks, but that point defines an important sector familiar to all those real or armchair sailors who listen to the European maritime weather forecasts.

4. In thermohaline, the *haline* means salt in Greek, and indeed the chloride of sodium chloride is only one of the group of elements called *halogens* (others include bromine, fluorine, and iodine).

5. Recent estimates (Levitus et al. 2000) based on direct sea water temperature measurements (i.e., with thermometers) show "warming of the world ocean" since the 1950s at least.

6. I put quotes around "cold" used as a noun because, for physicists, "cold" only means less heat (measured, for example, in calories or BTUs, or heat energy measured in joules), with everything above a temperature of absolute zero containing some heat energy. The southward-drifting cold waters contain less heat energy per cubic meter than the northward-drifting warm waters, and in that sense they transport "cold."

7. Simple as they may be, these models consist in applying the well-established laws of physics to the complex interactions of ocean and atmosphere. These laws include conservation of matter (in particular water and salt) and energy, and Newton's laws of motion. It seems that there are different ways in which deep ocean circulation can operate, some with the conveyor-belt transport of heat, some without, and that the transition from one mode of operation to the other can be quite rapid, with our present quite efficient conveyor belt close to the limit of instability. According to one calculation, the Great Salinity Anomaly of the 1970s took the Atlantic half of the way in the direction of conveyor shutdown.

8. I refer here to the energy and environment policies of the Bush administration, installed in 2001, reneging on environmental commitments and encouraging waste of coal and oil, while at the same time proposing significant cuts in environmental research. If the message of science is bad for profits, shoot the messenger.

9. CLOUDS, RAIN, AND ANGRY SKIES

1. From Shakespeare's *Antony and Cleopatra*, act 4, scene 14.

2. For a well-organized and complete description of the different cloud types, illustrated by a beautiful set of photographs, from below, within, and above the clouds (including from satellites), see Scorer and Verkaik (1989). The invention of the nomenclature is recounted by Hamblyn (2001).

3. The tropopause is the level separating the troposphere—i.e., the lower atmosphere, where temperatures generally decrease with altitude and where most weather phenomena take place (see fig. 7.1)—from the stratosphere, where temperatures increase upward. Tropopause altitude generally increases with atmospheric temperatures, and can be as high as 18 km (59,000 ft.) above the equator, as low as 10 km (33,000 ft.) at high latitudes in winter. But at high latitudes, cumulonimbus clouds form mostly in summer.

4. Haloes and other optical phenomena of the atmosphere have been described and illustrated by Minnaert (1954) as well as Scorer and Verkaik (1989).

5. Indeed, recent research suggests that in addition to producing easily detected shiptracks, maritime traffic has had a significant effect on cloud brightness over the North Atlantic as a whole. Cloud brightness is only part of the story. Working with satellite and aircraft data at the Hebrew University of Jerusalem and

at NASA's Goddard Space Flight Center in Greenbelt, Maryland, Daniel Rosenfeld (Rosenfeld and Lensky 1998) argues that smoke and urban pollution also significantly affect convective clouds and rainfall. More numerous but smaller droplets make clouds brighter, but they may reduce precipitation in the polluted areas.

6. Not all aerosols brighten clouds. Dark particles such as soot, related to inefficient industrial or biomass burning, reduce cloud brightness and warm the cloud layer, but at the same time they block sunlight to the surface. Enormous pollution of this kind, coming from south Asia, was observed during the international Indian Ocean Experiment (INDOEX) campaign in early 1999.

7. The question has not yet been answered whether contrails have (or can have in coming decades) a detectable effect on climate on the global scale. There are some indications that cirrus cloud cover has increased in the major air traffic corridors. Another debate is whether cosmic-ray particles arriving in the upper atmosphere, in greater or lesser numbers depending on solar activity, play any role in formation of cirrus.

8. Long a puzzle, ball lightning appears to result from fairly slow burning of a balloon-like structure a couple of feet in diameter, composed of vaporized carbon and silicon and formed after lightning heats the ground above 1,000°C, according to two scientists from New Zealand. Sprites are beautifully colored electrical discharges above clouds, observed from manned spacecraft and occasionally from aircraft, but not from the ground.

9. These are sea-level atmospheric pressures (the weight of the air above a unit area), with standard pressure (one "atmosphere") being 1,033 millibars (actually, the official international system would have us say 1,033 hectopascals) or 14.7 pounds per square inch. Many people are more familiar with the analogous readings in a mercury barometer corresponding to the height of the column of mercury (a very dense but liquid metal) that atmospheric pressure can support: 76 cm or 29.9 inches.

10. Both British and French weather services, as well as the European Centre for Medium-Range Weather Forecasting, did predict a powerful storm with strong winds, but the U.K.'s Met Office underestimated how bad it would be, and the seriousness of the risk did not get across to the British public, resulting in unnecessary injuries and loss of life. The circumstances were the subject of an investigation and reform. That storm was (luckily) well forecast in France.

11. Satellites in circular orbit above the equator at a distance of 36,000 km (22,600 mi.) go around the Earth in 24 hours, and so since the planet makes a full rotation in the same time, the satellite appears stationary, allowing continuous observation of the same real estate. The U.S. National Oceanic and Atmospheric Administration (NOAA) has the responsibility of maintaining two such Geostationary Operational Environmental Satellites in operation: GOES-East, for eastern North America as well as South America, much of the Atlantic Ocean as well as the eastern South Pacific; and GOES-West, for western North America and the eastern Pacific Ocean. The Japanese Meteorological Agency has consistently maintained

coverage of eastern Asia and the western Pacific. The European Meteorological Satellite consortium (Eumetsat) maintains Meteosat on the Greenwich meridian, observing Europe and Africa; Eumetsat loaned a spare Meteosat to the United States between 1992 and 1995, when premature GOES failures and delays in replacement would have left the U.S. with a single satellite, not enough to monitor both Pacific and Atlantic storms, and also provided a spare to support the Indian Ocean research campaigns (INDOEX) from 1998 to 2000. The United States moved a spare GOES to cover the Indian Ocean position as part of an earlier (global) campaign in 1979. That sector has always raised problems, the USSR (later Russia) never having got a working satellite up in position there, China also having had little success, and India having provided hardly any data to scientists even when its satellite worked.

12. In his book *Isaac's Storm*, Erik Larson (1999) tells the terrible story, with a critical perspective on the responsibilities of Isaac Monroe Cline and the U.S. Weather Bureau in the inadequate forecast of the hurricane's threat to Galveston. Fairly accurate warnings from Cuba were ignored. Cornelia Dean (1999:1–14) also describes the Galveston disaster, and the lessons that it should teach us for the future.

13. As noted earlier, under some conditions additional particles may lead to the formation of additional very small water droplets, too small to fall as rain, so that cloud seeding would be counterproductive. This would certainly confuse the statistical studies.

14. Smith (1872:413). Smith also writes: "A few centuries back we, perhaps, had arrived at such progress in filthiness of habits that we attained the dignity of producing epidemics amongst ourselves."

15. In the United States, the Clean Air Act as amended in 1990 was a major factor in the reduction of SO_2 emissions, with the institution of tradable emissions rights helping to make the cost of such "scrubbing" much less than was originally feared by industry lobbyists. Similar strong measures to reduce emissions were taken in the European Economic Community (now the European Union). The results are now measured in a reduction of rainfall, stream, and lake acidity in many areas of Europe and North America.

10. EARTH'S WATER, BETWEEN SKY AND SEA

1. *Sahel* is Arabic for *shore*, but it is used to denote the limit of the desert (a navigable expanse for caravans) as well as for the shore of the sea.

2. Many scientific terms for soils are in Russian because modern soil science—*pedology*—was founded in Russia, notably by V. V. Dokuchaev, in the 1870s. See also Zachar (1982).

3. Although often an oasis is thought of as a gift of nature, the people of the desert learned long ago that sometimes they could force Mother Nature to do more, when the wells they dug provided a continual (but not inexhaustible!) flow

supporting vegetation as well as drinking water for them and their herds. In the long term, the water will run out.

4. Bernard Palissy (1510–1590) made the case for this in 1580 in his "Admirable discourse on water and fountains." However, he is perhaps better known for his work in ceramics and decorative art. He narrowly escaped the anti-Protestant massacre of Saint Barthélemy in 1572, but after having been accused of "heresy" and sentenced in 1588 to be hanged and burned, he died in prison at the Bastille. In the seventeenth century, Edme Marriott (1620–1684) and Pierre Perrault (1611–1680; one of seven brothers active in areas ranging from architecture to finance, science, and writing fairy tales) proved the case by experimental research in the Paris basin, measuring in particular the flow in the Seine River. In the twentieth century, isotopic analyses have yielded much more complete data on the time that rainwater spends underground.

5. Reminder: a cubic kilometer (a billion cubic meters) of water has a mass of a billion metric tons. Another way of putting it: 80,000 cu. km of water on 8 million square kilometers of land (the area of the 48 conterminous states of the U.S.) represents a layer of water 10 meters (33 ft.) thick.

6. Wei, Ledoux, and de Marsily (1990) have developed a model of this regional underground flow.

7. See Craig M. Bethe and Stephen Marshak, "Brine Migrations Across North America—The Plate Tectonics of Groundwater," in Dietrich and Sposito (1997:253–81).

8. See Bonell (1993).

9. Continuous study of precipitation chemistry and how it affects soil and forest has been carried out at the Hubbard Brook Experimental Forest in New Hampshire since the 1960s (Likens 1999).

10. For Steven Pinker (1995:64), this is a myth. In his view, which he illustrates by quoting some of the terms used by Boston TV meteorologists, English has as many words (about a dozen) for snow as has Inuit or Yupik.

11. The word *levee* (or French *levée*), a raised river embankment, comes from the French-speaking colonists of Louisiana describing the banks of the lower Mississippi.

12. The Colorado State University Flash Flood Laboratory writes: "In the Eastern U.S. the flash floods associated with hurricane landfalls can offer up to 12 hours of lead-time. In the West, flash floods can offer less than one-hour advance warning. In arid and semiarid environments, steep topography, little vegetation, and infrequent precipitation in the form of intense thunderstorms typify the flash flood hazard areas of the Western United States."

13. The Field Museum expedition's findings are reported in Nials et al. (1978).

14. One might ask whether it isn't *excessively* watered, with 5,000 to 10,000 mm (200 to 400 in.) of rain per year. In fact, the situation is in a sense worse because most of that abundant rain falls in only a few months, with dryness dominating the rest of the year.

15. At the time, much was made of this drought as one of the signs of "global warming." But in the spring of 2001, the flooding in northern France due to overabundant rains in 2000–2001 was also being attributed to "global warming." Is all this just the usual media exaggeration? Some serious scientists believe that the climate changes summarized under the slogan of "global warming" may indeed involve increasing variability, i.e., *both* droughts and floods alternating, with increasing severity, at least in some parts of the world.

16. The twin GRACE satellites were successfully launched on March 17, 2002, from Plesetsk cosmodrome (spaceport and ICBM base) in northern Russia.

11. WATER AND MAN'S RISE TO CIVILIZATION

1. In the Scottish highlands and even more in the Alps, ecotype islands subsist, composed of plant, animal, and bird species that ranged over large areas of Europe when conditions were colder but find themselves confined to higher altitudes since the climate warmed. This process continues, reinforced by the recent warming, but also certainly influenced by the increasing afflux of city people with their heated ski lodges.

2. Certainly the corn that European colonists found on their arrival in the Americas had already dramatically changed from the earliest natural corn that existed. Genetic modification by breeding and selection is ancient, but opponents to modern genetic modification argue that we still don't know what we're doing. Nor will we learn more if extremists who vandalize research farms in France and elsewhere have their way.

3. Jared Diamond (1991:286–303) cites, among other cases, the extinction of the moa after the colonization of New Zealand by the Maoris, the disappearance of many native bird species on Hawaii, and the desolation on Easter Island.

4. The tree-ring results check with what has been directly measured since 1896. The study is reported by Stahle et al. (1998).

5. Read Jared Diamond's books (1991, 1997), among others. An interesting perspective on the development of civilization in the Americas is given by Peter Farb in his 1968 book *Man's Rise to Civilization*.

6. Joseph's interpretation of Pharaoh's dreams (Genesis 41:15–36, 41–57) led to Pharaoh adopting stockpiling of grain to prepare for years of drought. Steve Schneider and L. E. Mesirow gave the name *The Genesis Strategy* to their influential 1976 book on the importance of preparing for climate change.

7. Rosetta is where the famous Rosetta Stone was found, with a decree of Ptolemy V (196 B.C.) engraved in Greek, demotic, and in hieroglyphs, making it possible for Jean-François Champollion (1790–1832) to decipher the hieroglyphs and the ancient Egyptian language.

8. In his 1993 book *L'Homme et le Climat* (Man and climate), Jacques Labeyrie (pioneer in using radioactive isotopes to study ancient climates) noted

that two relatively rapid oscillations of two to three meters of sea level occurred in the three thousand years following the maximum of 5400 B.P. (3450 B.C.), and has hypothesized that this may explain the story of the Flood in Sumerian legends as well as in Genesis. However, as noted earlier, Ryan and Pitman (2000) set the Flood in what is now the Black Sea, but to them too it was in part a consequence of the 130-meter rise in sea level.

9. Camilla Calhoun reminds us of these figures in her fascinating "Perspective on New York City's Water supply," to be found at www.westchesterlandtrust.org/watershed/olive.htm.

10. Robert Koch (1843–1910), a pioneer of medical bacteriology, started his career as a country doctor in Germany. He became world famous in 1882 when he discovered the bacterium responsible for tuberculosis; he discovered the cholera vibrion the following year, and he made major contributions to the fight against plague, typhoid fever, and malaria, as well as tuberculosis and cholera. He was awarded the 1905 Nobel Prize for physiology and medicine.

11. In a Hoover Institution essay, Thomas Gale Moore (2000) describes the progress of the epidemic, which started from Chancay, Peru (near Lima), in January 1991 and spread to many other Latin American countries in following years, causing over a million cases and 11,000 deaths. He cites Anderson's (1991) report that the major factor for the epidemic's spread was Peru's decision to halt chlorination. Note, however, that Rita Colwell (1996) has explored links between El Niño events and cholera outbreaks.

12. The case against chlorine has been argued forcefully by biologist and former Greenpeace scientist Joe Thornton (2000) in his book *Pandora's Poison: Chlorine, Health, and a New Environmental Strategy*, enthusiastically reviewed by Terry Collins (*Nature*, July 6, 2000), but criticized by Ferdinand Engelbeen (*Nature*, September 28, 2000). Collins mentions a defense of chlorine in Howlett (1997). Many or most public health professionals consider that abandonment of chlorination of drinking water would be an unmitigated disaster (see note 11 above). As for THMs, although the Environmental Protection Agency (EPA) has not established a maximum contaminant level (MCL), the New York State Department of Health's MCL is 100 micrograms per liter, a value exceeded on occasion in some parts of the New York City water supply system, but not on average in 1999. Concern remains regarding contamination by various microbes, in particular *Giardia* and *Cryptosporidium*. The EPA has judged that water coming from New York's Croton watershed (supplying 10% of the city's water on average, but 30% during droughts) may not meet health standards because of bacterial contamination. However, the project to build a new water treatment plant in the city's Van Cortlandt Park (Bronx) appears blocked by community groups and a state court decision (*New York Times*, February 9, 2001); the city is liable for fines retroactive to 1999, and its water quality report must include the statement: "*Inadequately treated water may contain disease-causing organisms. These organisms include bacteria, viruses, and parasites, which can cause symptoms such as nausea, cramps, diarrhea, and associated headaches*" (emphasis in original).

13. La Butte aux Cailles ("quail butte") is today a lively neighborhood of Left Bank Paris, where you'll find quail at some butcher shops and on restaurant tables, but nowhere else. The artesian well supplies warm water to a large public swimming pool. The butte also was the place where Pilâtre de Rozier and the marquis d'Arlandy landed at the end of the first hot-air balloon (*montgolfière*) flight on November 21, 1783. The start of that flight, to the west in Passy (now also part of Paris), was witnessed by Benjamin Franklin as well as the Montgolfier brothers.

14. The Ecole Polytechnique (polytechnic school) was founded in 1794 to train scientists and engineers for public service to the young Republic of France. Having taken the title of emperor, Napoleon put the Ecole under military control with the mission of training engineers for his army. Although administratively it remains a military school, today's Ecole should be compared to MIT rather than West Point. Its students, selected after fierce competition, actually pay no tuition; instead, they receive a "presalary" but owe ten years' service to the State unless they or a prospective private employer buy back the "debt." The Ecole trains engineers and scientists, many of whom go on to top-level public or corporate administration. Originally located in the Latin Quarter of Paris (so-called because students from all over Europe worked and conversed there in Latin in the Middle Ages), the school outgrew its quarters and moved in 1974 to a suburban site in Palaiseau, 25 km (15 mi.) to the south. Several research laboratories (including the one where I work) were installed on the same site.

15. The flood level of 1910, marked on many buildings in the historic center of Paris, is hip-high in many places, even though the street level is at least one story above the usual Seine level. The banks of the Seine (and the express roadways unfortunately built there in the 1960s) are often flooded at the end of the winter, but the water has not reached street level since 1910.

16. Gerard Koeppel (2000) has written an engaging history of New York City's water supply, also summarized briefly in the National Research Council's (NRC) assessment of the city's strategy (O'Melia 1999). See also note 9 above.

17. The NRC's assessment (O'Melia 1999) also notes that "while New York City has retained control over its system, and has encountered much anti-City hostility in its source regions, the metropolitan Boston system has long been administered by agencies serving a region containing more than half the state's population and therefore has gained broader political support in its efforts to protect its water supplies." In particular, "New York City has had to engage in lengthy and delicate negotiations with upstate interests."

18. However, growing awareness of real (or imaginary) risks related to the presence of small amounts of organochlorines in drinking water has led to tighter standards (see note 12 above) and an escalation of requirements. What, after all, is "reasonably" safe? In Paris, new standards on the allowable amount of lead will lead to expensive replacement of plumbing (the word comes from the Latin and French words for lead, long the material of choice for such applications) in buildings, at owners' and tenants' cost; alternatively, systems for removing the lead in the water at the supply to individual apartments may turn out to be reliable and cheaper.

19. In the United States the Bangladesh crisis was reported in the *New York Times* on November 10, 1998, as what could be "the biggest mass poisoning in history." Many additional reports have appeared in the British press, both scientific and lay (Nickson et al. 1998; Pearce 2001), and of course in Bangladesh and India. Ongoing developments can be followed at the Internet site www.bicn.com/acic.

For the United States, the EPA's interim maximum contaminant level (MCL) for arsenic in drinking water used to be 50 micrograms per liter, five times higher than the World Health Organization's recommendation. In 1996, Congress required the EPA to propose a more stringent standard in 2000, and EPA proposed adoption of the 10 micrograms per liter limit (EPA 2001). Early in 2001, the Bush administration sparked lively controversy when it announced its intention *not* to lower the allowed arsenic level in drinking water supplies; arsenic being well-known as a poison, this position is hard to defend, even in the 2001 Republican-controlled House of Representatives. Dangers of long-term exposure to such low (not Bangladeshi) levels of arsenic are discussed in a report by the National Academy of Science's Commission on Life Sciences (1999).

20. Water experts Peter Gleick, Sandra Postel, and Diane Martindale discuss these issues in several articles in the February 2001 issue of *Scientific American*.

21. It may be remarked that all agricultural exports involve in effect the export of water, and that apparently profitable Israeli citrus exports depend on strong government subsidy of the water infrastructure and supply for irrigation. There must be more real profit in exporting high technology. See also Plaut (2000).

22. According to Peter Gleick (note 20 above), there are more than 70,000 dams in the United States alone.

23. René Dubos (1901–1982), though born and trained as an agronomist in France, made nearly all his career at the Rockefeller Institute in New York. Discoverer in 1940 of the first antibiotic (gramicidin) used in medicine, he made important contributions to biology and medicine and later in raising awareness of environmental problems (Dubos 1970, 1980). He was the first to use the slogan "Think globally, act locally."

24. News on these techniques is given in the magazine *Agroforestry*, published by the International Centre for Research in Agroforestry (ICRAF), based in Nairobi, Kenya.

25. Spraying (though not with DDT) has made a comeback with the appearance in 1999 of West Nile encephalitis in the United States and Canada, with many birds and some horses and humans as victims. The virus is carried by *Culex* mosquitoes, and the infection probably reached America by airplane, as the disease first appeared in the vicinity of international airports in the New York area. Similarly, the *Aedes albopictus* (Asian tiger) mosquito, which can carry dengue fever and is suspected of carrying eastern equine encephalitis, has been extending its range in southern Europe, and may soon reach France. In the United States, it was identified in Houston, Texas, in 1987 had has since reached southern New Jersey.

26. For a vivid description of the frozen pond's beauty, and the full story of the ice harvest, see Thoreau (1897: ch. 16, "The Pond in Winter"). Through his book *Walden* (1854), Henry David Thoreau (1817–1862) profoundly influenced the beginnings of the American environmentalist movement. In addition, his 1849 essay *On the Duty of Civil Disobedience* inspired Gandhi and his organization of civil disobedience first in South Africa, later in India.

27. Zbigniew Jaworowski (1999) of the Central Laboratory for Radiological Protection (Warsaw, Poland) argued that even in the case of the Chernobyl accident, damage to health has been grossly exaggerated, where the low radiation doses associated with fallout (not the high doses of the "liquidators") are concerned. Strong letters of objection and Jaworowski's vigorous rebuttal appeared in the April and May 2000 issues of *Physics Today*. The debate continues on whether or not thresholds limit deleterious health effects of low radiation doses.

28. I prefer to write Medvedev's given name as Jaurès because it honors French pre-World War I socialist leader Jean Jaurès and becomes Zhores only after two-way transliteration into and out of the Cyrillic alphabet.

29. Medvedev's 1977 book *Nuclear Disaster in the Urals* remains an excellent introduction, especially in the newer editions including extracts from CIA archives, declassified thanks to the Freedom of Information Act. Bradley, Frank, and Mikerin (1996) have discussed nuclear contamination from both Soviet and American weapons complexes, with a fairly detailed map of the Mayak area where the 1957 disaster took place and where radioactive waste discharge was routine since 1949. They also discuss the even larger routine discharges into surface water and underground from the Tomsk-7 and Krasnoyarsk-26 complexes in Siberia.

30. For physicists' perspectives on these issues, see the June 1997 special issue of *Physics Today* on radioactive waste.

12. PROBLEMS FOR THE 21ST CENTURY

1. Most scholars agree, however, that the count is wrong, and that Jesus of Nazareth was born near Bethlehem a few years before 1 A.D. See Stephen Gould's *Questioning the Millennium* (1997) for a discussion of calendar problems.

2. Compact and readable discussions of world population history are by Cipolla (1978) and by McEvedy and Jones (1978). The very complete and readable book by Joel Cohen (1995)—*How Many People Can the Earth Support?*—has strongly influenced my own thinking developed here. Robert W. Kates also discusses the issue in Ausubel and Longford (1997:33–55). A very recent paper by Lutz, Sanderson, and Scherbov (2001) is explicitly titled, "The End of World Population Growth."

3. The dietary Calories with which most people are familiar are in fact kilocalories, one calorie being the quantity of heat needed to raise the temperature of a gram of water by one degree Centigrade. A kilocalorie or Calorie raises the temperature of a kilogram of water by the same amount.

4. The argument that human activities now dominate both freshwater and the biosphere is made by Postel, Daily, and Ehrlich (1996) and Vitousek et al. (1997).

5. In one of his 504 *Maximes* (from one-liners to long paragraphs) on all sorts of subjects, François de La Rochefoucauld (1613–1680) wrote that "hypocrisy is the tribute that vice pays to virtue." It seems to me that some though certainly not all of the complicated arguments in defense of the environment, invoking the economical value of the services that certain ecosystems render, constitute a tribute that virtue is paying to vice (or to the "dismal science" of economics). Of course, this may also be viewed as more or less effective political tactics, and I may be naïve in seeing this tendency as an unnecessarily cynical view of what moves my fellow citizens of the planet. That could bring us into the debate about "sustainable development" and whether such a term has signification or is simply a shibboleth. We could, especially sitting around a jug of good wine, also get into the discussion of the value of an environmental ethic for the survival of *Homo sapiens*, just as there has been discussion concerning the origins of altruism, a reaction that improves the chances of the species in the arena of evolution by way of natural selection, although it often leads to the death of the individual. See also Wilson (1998).

6. Rain, water from heaven, is indeed the gift of God in the Old Testament, at least for those who "hearken unto the voice of the Lord," to whom "the Lord will open . . . His good treasure the heaven to give the rain of thy land in its season, and to bless all the work of thy hand" (Deuteronomy 28:1 and 12). Note that the seasonal rhythm is respected, and that humans must supply work. Withholding of rain figures as punishment (Deut. 28:24).

In the Koran (Qur'an), too, lifegiving rain, is indeed a gift of God. For Islamic law, or *shari'a*, water is first of all a free commodity. At the same time, its nature as a value-added commodity is recognized. Prior appropriation is legitimate, but any surplus must be distributed, and the owner is liable for misuse. See Dolatyar and Gray (2000:172).

7. These problems are treated in many different ways by different legal systems, even by different states of the U.S. Federal law applies to interstate waterways (for example, the Ohio, Mississippi, and Missouri Rivers), and international treaties govern rivers that flow through or between different countries (for example, the Rhine and Danube Rivers in Europe). Some countries insist on the distinction between international or common waters of rivers that form a boundary between two countries or states, and transboundary waters of rivers that cross a boundary flowing from one country or state to another. But many rivers are both: the Colorado River defines part of the border between Arizona and California and Nevada, but its waters come mostly from Colorado by way of Utah, and a residue flows through Mexico. The Rio Grande (or Rio Bravo del Norte) defines the Texas-Mexico border, but comes from New Mexico and Colorado. The U.S.-Mexican water treaty of 1944 includes the statement that the treaty "excludes the idea of a legal obligation" (G. Rowley, in Amery and Wolf 2000:233). Where Syria and Iraq consider the Tigris and Euphrates as international waters to be shared, Turkey in-

sists on their transboundary character, and rather than accepting an obligation to share, it proposes "equitable, reasonable, and optimal utilization."

8. The Austrian and Ottoman Empires signed a treaty concerning the Danube as early as 1619, and France and Germany codified navigation rights on the Rhine in a 1697 treaty. Vlachos (1990:186) notes that "by the early 1970s, about 286 international treaties concerning water resources had been signed." See also Glantz (1994:88–112) for brief discussions of the role of regional organizations.

9. Syed Kirmani discusses the Indus and Mekong River basins in Vlachos (1990:200–205); Richard Kattelmann (Vlachos 1990:189–94) examines the need for cooperation between India and Nepal for better flood management in the Ganges plain.

10. James London and Harry Miley Jr. discuss this in their article (in Vlachos 1990:231–35) on "The Interbasin Transfer of Water: An Issue of Efficiency and Equity," mostly on the basis of U.S. experience. They note that the states of Washington and Louisiana, "discussed as potential sources of water supply for the states of California and Texas, respectively," have adopted restrictive instream flow provisions "targeted more at external threats than at instate conflicts."

11. This is a rough translation of what Claude Allègre (1990) wrote in his book *Economiser la Planète* (Economizing the planet).

12. Smith and Al-Rawahy (Vlachos 1990:217–22) consider that even without specific projects for the Blue Nile, the increasing population and water needs of Egypt are already creating a crisis with potential for conflict.

13. The Turkish acronym GAP stands for Guneydogu Anadolu Projesi, or "southeast Anatolia project." Some 90 percent of the annual mean Euphrates River flow originates in Turkey, and 38 percent of the Tigris. President Türgut Özal of Turkey did propose construction of a "peace water pipeline," in particular to satisfy Syrian needs, but this was regarded by Syria as proof of perfidious Turkish intentions to restore the Ottoman Empire. See Tekeli (Vlachos 1990:206–16) and Starr (1995) as well as Berkoff (1994) and Amery and Wolf (2000).

14. Following the defeat of the Ottoman Empire in World War I, the 1920 Treaty of Sèvres promised autonomy both to Armenians and to Kurds, but the proposed U.S. mandate for Armenia in northeast Anatolia (Lake Van, Mount Ararat) never came about, and with the strong Turkish rebound under Gen. Mustafa Kemal (later Atatürk), the French withdrew their troops from Cilicia (southeast Anatolia). Despite vigorous resistance, Armenian survivors of the 1915 genocide were able to maintain themselves in the area around Erevan only as part of the Soviet system, and recovered independence only after the breakup of the USSR in 1991. As for Kurdistan, strong Turkish resistance, and British and French reluctance to see the oil-rich territories around Mosul in northern Iraq go to a Kurdish state, led to the abandonment of the project.

15. A sizable "neutral territory" was defined in the 1920s between Saudi Arabia and Iraq, just west of Kuwait, in order to give equal access to wells (of water!) to Bedouin tribesmen coming from any direction. Some of the borders between

Saudi Arabia and its neighbors remained undefined until the last quarter of the twentieth century.

16. Depending on which etymology you prefer, Beersheba can mean the well of seven lambs, or the well of the oath, or the well of good fortune (Hertz 1967).

17. When Moses presents the full set of (613) commandments to the children of Israel, he begins the description of the Promised Land with: "For the Lord thy God bringeth thee into a good land, a land of brooks and water, of fountains and depths, springing forth in valleys and hills." (Deuteronomy 8:7).

18. The Siloam Pool, located within the walls of the City of David, lies outside the walls of the present "old city," whose plan corresponds to the period of the Second Temple and later. The spring gushes water for half an hour every four hours. An earlier open tunnel existed in the time of King Solomon (965–928 B.C.). Construction of Hezekiah's tunnel is mentioned in (2 Chronicles 32:2–4 and 32:30 and in 2 Kings 20:20). The Siloam Inscription describing the piercing of the tunnel was discovered in 1880 and can be viewed in the Archaeological Museum of Istanbul; a copy exists in Paris.

19. This did not protect the other towns of Judah, many of which were destroyed by Sennacherib's army. Also, Hezekiah still ended up paying tribute to Sennacherib, who claimed victory. See Tadmor (in Ben-Sasson 1976:139–46), and also Issar (1990). Note that the prophet Isaiah decried Hezekiah's temporary alliance with Babylon against Assyria (2 Kings 18–20; Isaiah 36–37).

20. Jesus cured a man blind since birth (not "visually challenged," blind!), sending him to the Siloam Pool (John 9:6–7).

21. From 1949 to 1967, some of the eastern shore of Kinneret, contested by Israel and Syria, was a demilitarized zone. After the Six-Day War, the Israeli Defense Force occupied this territory as well as the Golan Heights.

22. In a sense, this constituted partial unilateral implementation by Israel of the 1948 Lowdermilk plan, which envisaged using practically all the Jordan River's flow into the salty Dead Sea (396 meters below sea level) for irrigation, replacing it by water from the Mediterranean and using the altitude difference to generate electric power. Getting freshwater from Kinneret (normally at 209 meters below sea level) to the National Water Carrier (NWC) requires power to pump it to an altitude 151 meters above sea level. When rainfall over the upper Jordan basin is low, as in recent years, excessive pumping can degrade water quality in Kinneret, the lower Jordan River, and underground. Nevertheless, by 2001 the Israel Hydrological Service lowered the "red line" in Kinneret 30 cm below the already record 214 meters below sea level (David Rudge in the *Jerusalem Post*, March 6, 2001). Note that in parallel with Israel's NWC, the Kingdom of Jordan constructed the King Abdullah Canal, diverting water from the Yarmuk River upstream of its confluence into the Jordan, using it for irrigation east of the Jordan River and restoring the residual water to the lower Jordan above the Dead Sea. Israel's NWC removes water from the Jordan basin. Beginning with the Lowdermilk plan, and the later and somewhat different Hayes and Johnston plans, the international community and

the United States have repeatedly sought to promote international cooperation in using water in this water-scarce region.

23. Then Israeli Defense Minister (now Prime Minister) Ariel Sharon has been characterized as a "water hawk," advocating transfer of Litani River water to the Jordan. However, in a 1991 interview, quoted by Woolf (1995; ch. 2), also by Amery and Wolf (2000), Israeli Gen. Avraham Tamir declared: "Why go to war over water? For the price of one week's fighting, you could build five desalination plants. No loss of life, no international pressure, and a reliable supply you don't have to defend in hostile territory" (see also Tamir 1998).

24. Drawing the exact northern borders of Palestine in 1919 involved negotiation between Great Britain as mandatory power for Palestine on the one hand, and France for Syria and Lebanon on the other. Arab Palestinians and Jewish immigrants had little voice on the matter, although it is true that Chaim Weizmann (thirty years later the first president of Israel) had expressed to the British the desirability of including important water sources within the northern border. Those ambitions were only partly satisfied.

25. This quote was translated from the Israeli newspaper *Ha'aretz* (January 1, 1993) by the PEACE Middle East Dialogue Group (www.ariga.com/peacewatch). Issar returned to the question in June 1998, noting the interrelated issues of Jewish settlements, the convoluted Israel-Palestine border, Israeli security concerns and water resources, and concluding that "the debate slides into a well-known weary pattern."

26. Indeed, both the Israeli water commissioner and the companies making money from the operations claim significant success in increasing rainfall over Israel. However, most specialists are skeptical about "rainmaking," although it is conceivable that artificial introduction of cloud condensation nuclei (see ch. 9) can change the site of rainfall. However, given the small distances separating Israeli and Palestinian lands, and the uncertainties regarding where extra rain if any will actually fall, it is doubtful that the Palestinians have a serious grievance here.

27. None of this discussion appeared in the original French edition of this book, but I was asked to develop the subject in the English-language edition; I have done so with some trepidation. Every day (from autumn 2000 to mid-2002) the news is worse and worse, and I have found it surrealistic to read the often quite optimistic recent books and reports (Amery and Wolf 2000, Dolatyar and Gray 2000, Hillel 1994, Starr 1995, Tamir 1995, and Woolf 1995, as opposed to Chesnot 1993 and Clarke 1993).

13. BUTTERFLIES AND HUMANS IN A WARMING GREENHOUSE

1. This story, with its numerous scientific and political ramifications, has been described in several books, in particular by Roan (1990). Some illicit manufacture of CFCs has continued outside of North America and Europe since the phase-out

decided in the 1987 Montreal Protocol, but only in small amounts. A major challenge is the recovery and recycling of all the CFCs still in refrigeration and airconditioning equipment, before they escape into the atmosphere when such equipment is junked at the end of its lifetime.

2. Pierre Simon Laplace (1749–1827), named count by Napoleon in 1806, became a marquis after the Bourbon restoration in 1817. He made major contributions to mathematics and in particular probability theory and celestial mechanics. There is a Rue Laplace in Paris, and a Promontorium Laplace on the Moon.

3. During the Rio "Earth Summit" of 1992, a UN Framework Convention on Combating Climate Change was drafted and later signed by most governments; in the case of the United States, President Clinton signed the treaty, previous President Bush having refused signature. In order to translate that very vague accord into some modest but significant quantitative goals, governments of developed countries (including the U.S.) signed the Kyoto Protocol at the end of 1997. The agreement was to reduce greenhouse gas emissions by percentages ranging from 0 to 8 percent with respect to the 1990 emissions levels by the period 2008–2012. Agreed reductions were more for some European countries (12.5% for the United Kingdom and 21% for Germany) but less for others (8% for the European Union as a whole). The United States commitment was for 7 percent, that of Canada and Japan for 6 percent. Australia agreed only to limit its *increase* of emissions to 10 percent. Developing nations made no commitments, considering that their *per capita* emissions rates were (and still are) much lower than those of the rich countries. Hardly any country actually ratified the protocol before 2001, and in the U.S., after three years during which the elected Clinton-Gore administration did not dare present the protocol for ratification to the Senate, the newly installed conservative Bush administration declared its intention to *withdraw* the U.S. signature. Note that even if the Kyoto Protocol's provisions were to be fully respected, global greenhouse gas emissions would be reduced by very little (Bolin 1998). The significance of following through on the Kyoto Protocol would be that it would force technologically advanced countries to consider seriously how to maintain and develop the world's economy with less use of fossil fuels (Weizsäcker, Lovins, and Lovins 1997; Rosenfeld, Kaarsberg, and Romm 2000), and generally how to produce more real wealth with less energy consumption. Recent reports (Erik Eckholm in the *New York Times*, June 15, 2001) indicate that China has maintained rapid economic development while slowing its greenhouse gas emissions, i.e., keeping its per capita emissions at a very low level (0.6 metric tons carbon equivalent compared to 5.6 for the United States and 1.9 for France). Figures of the U.S. Department of Energy's Carbon Dioxide Information Analysis Center show that between 1990 and 1996, efficiency of energy use in China rose from 0.66 to 1.15 million dollars (M$) of gross domestic product (GDP) per thousand tons (kT) of emitted CO_2 (in the more technologically advanced United States, from 1.17 to 1.42 M$ GDP/kT CO_2; in nuclear-powered France from 2.87 to 3.50 M$ GDP/kT CO_2).

The countries of the European Union finally ratified the Kyoto Protocol in May 2002, as did Japan, and with ratification by fifty-five countries the protocol comes into effect *without* the United States.

4. For example, one model's prediction of rainfall and runoff changes indicates an easing of the water crisis in China's Yellow River (Huang He) basin, taking into account predicted population change, but making things worse along the Yangtze Kiang (Chang Jiang River) to the south. That model also has climate change providing more water in much of the western United States and Canada and part of Mexico, but reducing water availability in the eastern U.S. and central Canada. See Vörösmarty et al. (2000). Other models give partly contradictory results (IPCC 2001). Obviously more research is needed to see where we are going.

14. BACK TO THE ICE AGE

1. Indeed, in 1972, crop failures in the Soviet Union, coupled with a sharp El Niño–induced drop in the anchovy fishery off Peru, led to agricultural prices being sharply pushed up by sly Soviet grain purchases and the shortage of fishmeal for animal feed. The CIA (1974) commissioned a report on the possible geopolitical impact of future climate accidents of this type, a report eventually made public. In 1975 the (U.S.) National Defense University published an inquiry into the opinions of a number of climatologists regarding the likelihood of warming or cooling between then and the year 2000. One may note that the majority prognosis, a limited warming, turned out to be correct. The United States and other countries began organizing climate research programs at about the same time, and the first World Climate Conference took place in Geneva in 1979. It decided that the situation merited attention, but that it was too early to convene an international conference at the ministerial (cabinet secretary) level. That changed over the next decade, leading up to the 1992 Rio "Earth Summit."

2. This point is developed in a recent Club of Rome study (Weizsäcker, Lovins, and Lovins 1997). Even leaving aside issues of what constitutes a better quality of life, modern high-tech societies such as the United States (Rosenfeld, Kaarsberg, and Romm 2000) have already learned how to produce more and better material goods with less expenditure of energy. It is grotesque that the U.S. government, hijacked in 2000–2001 by the oil and coal industries, should maintain that the high-tech U.S. economy depends on wasting crude oil and coal. Short-term difficulties may arise, but even in the medium term, not to mention the long term, the United States and other advanced societies (Europe, Japan) have every interest to use the most advanced technologies available, and to burn less.

3. From A. G. Babayev, in Glantz (1977:205). Babayev did however admit that "desolation of extensive territories stemmed not so much from weather changes as from social processes."

4. The biblical injunction has been read by many environmentalists as the root of Western Man's original sin against the environment. Genesis 2:5 reads, however, "Be fruitful, and multiply, *and replenish the Earth*, and subdue it" (emphasis added). Mastery entails stewardship. One may note in any case that Eastern philosophy did not prevent the very early deforestation of China. Also, I read the Genesis phrase not so much as a command of God, whatever God may be, as a singularly prescient intuition of the destiny of human beings on this planet, considering that when those lines were written a few thousand years ago, most scientific discoveries and technological advances were still far in the future. But agriculture had begun.

5. This is in part the argument of Bill McKibben (1990) in his book *The End of Nature*, but I do not share all his sentiments. In particular, it seems to me that the development of human culture and agriculture put an end to a certain type of nature well before the twentieth century. However, general awareness of the fact that humans change the planet as a whole only came in the twentieth century.

15. CONCLUSION—THE END OF THE STORY?

1. Paris tap water is excellent, and safe; it has less of a chlorine taste than tap water in New York City or most other places in the United States (less use being made of chlorine and more of activated carbon). Some neighborhoods are still supplied with local well water rather than treated water from upstream Seine and Marne Rivers. But many Parisians prefer to buy bottled water and lug it around. And wine is still a drink of choice. The Paris Wine Museum (Musée du Vin) is located just at the end of the Rue des Eaux ("Street of the Waters"), in the 16th arrondissement. A spring was discovered there in 1645, giving the first of the Passy mineral waters; but it ran dry in 1770 (Hillairet 1963).

BIBLIOGRAPHY

The bibliography of the original edition of this book included many semipopular books in French. Here I list those English translations that exist, but I have kept the references to the French language editions for the others. For some of the books, I do not give the most recent edition but rather the one that I consulted. In general, and always where I give specific page references, I indicate the date of that edition (for most older works not the date of the original publication). I have added references to many publications that have appeared since the French edition of this work, as well as some semipopular books on topics covered in the French books in the list that have not been translated into English. I also have given a much more complete list of publications, some of them technical, on specific points. References are listed alphabetically by author's last name (not considering the French particle *de*), and in some cases by organization or acronym. The numbers in the text and tables of this book on water stocks, fluxes, and use come from the much more complete (but sometimes mutually inconsistent) tables found in Baumgartner and Reichel (1975), Berkoff (1994), Berner and Berner (1987), Cohen (1995), FAO (1997), Herschy and Fairbridge (1998), Issar (1990), and UNESCO (1978). Historical remarks in the text and notes are based on various sources, including in particular but not only Ben-Sasson (1976), Bensaude-Vincent and Stengers (1993), Bynum, Browne, and Porter (1983), Crombie (1959), Frisinger (1983), Hillairet (1963), Kutzbach (1979), McEvedy and Jones (1978), Salem (1990), and the (French) *Encyclopaedia Universalis*.

Allègre, Claude. 1988. *The Behavior of the Earth: Continental and Seafloor Mobility*. Cambridge: Harvard University Press.
——. 1990. *Economiser la Planète*. Paris: Fayard.

Amery, Hussein A. and Aaron T. Wolf, eds. 2000. *Water in the Middle East: A Geography of Peace*. Austin: University of Texas Press.

Anderson, C. 1991. Cholera epidemic traced to risk miscalculation. *Nature* 354: 255.

Arrhenius, Svante. 1896. On the influence of carbonic acid in the air upon the temperature of the ground. *Philosophical Magazine* 41: 237–76.

Ausubel, Jesse H. and H. Dale Langford, eds. 1997. *Technological Trajectories and the Human Environment*. Washington, D.C.: National Academy Press.

Ball, Philip. 1999. *H₂O: A Biography of Water*. London: Weidenfeld and Nicolson.

Baumgartner, A. and E. Reichel. 1975. *The World Water Balance*. Translated by R. Lee. Munich and Vienna: R. Oldenburg Verlag.

Ben-Sasson, H. H., ed. 1976. *A History of the Jewish People*. Cambridge: Harvard University Press.

Bensaude-Vincent, Bernadette and Isabelle Stengers. 1993. *Histoire de la Chimie*. Paris: La Découverte.

Berkoff, Jeremy. 1994. A strategy for managing water in the Middle East and North Africa. Washington, D.C.: World Bank.

Berner, E. K. and R. A. Berner. 1987. *The Global Water Cycle*. Englewood Cliffs, N.J.: Prentice-Hall.

Bernstein, M. P., S. A. Sanford, and L. J. Allamandola. 1999. Life's far-flung raw materials. *Scientific American* 281 (July): 26–33.

Berger, André. 1992. *Les Climats de la Terre*. Brussels: DeBoeck University.

Bolin, Bert. 1998. The Kyoto negotiations on climate change: A science perspective. *Nature* 279: 330–31.

Bonell, M. 1993. Progress in the understanding of runoff generation dynamics in forests. *Journal of Hydrology* 150: 217–75.

Brack, André and François Raulin. 1991. *L'évolution Chimique et les Origines de la Vie*. Paris: Masson.

Bradley, Don J., Clyde W. Frank, and Yevgeny Mikerin. 1996. Nuclear contamination from weapons complexes in the former Soviet Union and the United States. *Physics Today* 49.4 (April): 40–45.

Broecker, Wallace S. 1985. *How to Build a Habitable Planet*. Palisades, N.Y.: Eldigio Press.

——. 1997. Thermohaline circulation, the Achilles' heel of our climate system: Will man-made CO₂ upset the balance? *Science* 278: 1582–88.

Bryan, F. and A. Oort. 1984. Seasonal variation of the global water balance based on aerological data. *Journal of Geophysical Research* 89 (D7): 11,717–11,730.

Burroughs, William James. 1997. *Does the Weather Really Matter? The Social Implications of Climate Change*. Cambridge: Cambridge University Press.

Bynum, W. F., E. J. Browne, and Roy Porter, eds. 1983. *Macmillan Dictionary of the History of Science*. London: Macmillan.

Carr, Michael H. 1996. *Water on Mars*. New York: Oxford University Press.

Chesnot, Christian. 1993. *La Bataille de l'Eau au Proche-Orient*. Paris: l'Harmattan.

CIA (Central Intelligence Agency). 1974. *Potential Implications of Trends in World Population, Food Production, and Climate*. CIA, Directorate of Intelligence, Office of Political Research. OPR-401. Washington, D.C.: Library of Congress (DOCEX Project).

Cipolla, Carlo M. 1978. *The Economic History of World Population* (7th ed.). Hammondsworth and New York: Penguin.

Clarke, Robin. 1993. *Water—The International Crisis*. Cambridge: MIT Press.

Cohen, Joel E. 1995. *How Many People Can the Earth Support?* New York: Norton.

Colloque de la Villette. 1994. *Les Paradoxes de l'Environnement: Responsabilité des Scientifiques, Pouvoir des Citoyens*. Paris: Albin Michel.

Colwell, R. R., 1996. Global climate and infectious disease: The cholera paradigm. *Science* 274: 2025–31.

Commission on Life Sciences. 1999. *Arsenic in Drinking Water*. Washington, D.C.: National Academy Press.

Comte, Auguste. 1835. *Cours de Philosophie Positive*. Tome II: *La Philosophie Astronomique et la Philosophie de la Physique*. Paris: Bachelier.

Courel, Marie-Françoise. 1985. *Etude de l'évolution récente des milieux sahéliens à partir des mesures fournies par les satellites*. Paris: Centre scientifique d'IBM-France.

Crombie, Alistair C. 1959. *Medieval and Early Modern Science*. Vol. 1, *Science in the Middle Ages: V–XIII Centuries*. Garden City, N.Y.: Doubleday Anchor.

Danchin, Antoine 1990. *Une Aurore de Pierres—aux Origines de la Vie*. Paris: Seuil.

Darwin, Charles. 1958. *The Origin of Species* (1859). New York: Mentor.

Davies, Paul. 1993. *The Mind of God*. London: Penguin.

Dean, Cornelia. 1999. *Against the Tide: The Battle for America's Beaches*. New York: Columbia University Press.

Deléage, J.-P. 1997. L'eau: Une ressource à préserver. *La Science au Présent—1998*, 249–61. Paris: Encyclopaedia Universalis.

Demangeot, Jean. 1992. *Les Milieux " Naturels" du Globe*. Paris: Masson.

Deming, D. 1999. Recent advances in studies of the possible influence of extraterrestrial volatiles on Earth's climate and the origins of the oceans. *PALAEO* 146: 33–35.

Dessler, A. J. 1991. The small-comet hypothesis. *Reviews of Geophysics* 29: 355–82.

Diamond, Jared. 1991. *The Rise and Fall of the Third Chimpanzee*. New York: Vintage.

——. 1997. *Guns, Germs, and Steel: The Fates of Human Societies*. New York: Norton.

Dietrich, William E. and Garrison Sposito, eds. 1997. *Hydrologic Processes from Catchment to Continental Scales*. Palo Alto: Annual Reviews.

Dolatyar, Mostafa and Tim S. Gray. 2000. *Water Politics in the Middle East: A Context for Conflict or Cooperation?* New York: St. Martin's.

Dubos, René. 1970. *Reason Awake: Science for Man*. New York: Columbia University Press.

——. 1980. *The Wooing of Earth*. New York: Scribner's.

Duplessy, Jean-Claude. 1996. *Quand l'Océan se Fâche: Histoire Naturelle du Climat.* Paris: Ed. Odile Jacob.

Ehrlich, Paul R., Anne H. Ehrlich, and John P. Holdren. 1977. *Ecoscience: Population, Resources, Environment.* San Francisco: W. H. Freeman.

Engelbeen, Ferdinand. 2000. Don't dismiss chlorine; it could help us to avoid the fate of the Romans. *Nature* 407: 445.

EPA (Environmental Protection Agency). 2001. National Primary Drinking Water Regulations; Arsenic and Clarifications to Compliance and New Source Contaminants Monitoring. *Federal Register* 66.14: 6976–7066.

Erwin, Douglas H. 1993. *The Great Paleozoic Crisis: Life and Death in the Permian.* New York: Columbia University Press.

——. 1996. The mother of mass extinctions. *Scientific American* 275.1: 56–62.

FAO (Food and Agriculture Organization). 1997. *Irrigation in the Near East Region in Figures.* Water Report 9: 1020–1203. Rome: FAO of the United Nations.

Farb, Peter. 1968. *Man's Rise to Civilization.* New York: Avon.

Frank, L. A. and J. B. Sigwarth. 1999. Comprehensive investigation of atmospheric holes with the *Polar* spacecraft that eliminates the possibility of camera errors and firmly establishes the reality of atmospheric holes and their geophysical effects. *Journal of Geophysical Research* 104: 115–39.

Frisinger, H. Howard. 1983. *The History of Meteorology: To 1800.* Boston: American Meteorological Society.

Gibbs, W. W. 1998. Endangered—Other explanations now appear more likely than Martian bacteria. *Scientific American* 278.4: 13–14.

Gibrat, Robert. 1966. *L'Energie des Marées.* Paris: Presses Universitaires de France.

Glantz, Michael H., ed. 1977. *Desertification: Environmental Degradation in and Around Arid Lands.* Boulder: Westview.

——, ed. 1994. *The Role of Regional Organizations in the Context of Climate Change.* NATO ASI series I: Global Environmental Change, vol. 14. Berlin: Springer-Verlag.

Goethe, Johann Wolfgang von. 1887. *Faust* (Gesamtausgabe textrev. H. G. Gräf). Leipzig: Insel-Verlag. English translation, B. Taylor. 1912. New York: Modern Library.

Gold, Thomas. 2000. *The Deep Hot Biosphere.* New York: Copernicus.

Goldenberg, Stanley B., Christopher W. Landsea, Alberto M. Mestas-Nuñez, and William M. Gray. 2001. The recent increase in Atlantic hurricane activity: Causes and implications. *Science* 293: 474–79.

González, F. I. 1999. Tsunami! *Scientific American* 280.1: 44–55.

Goudie, Andrew S. 1977. *Environmental Change.* Oxford: Clarendon Press.

Gould, Stephen Jay. 1991. *Wonderful Life: The Burgess Shale and the Nature of History.* London: Penguin.

——. 1997. *Questioning the Millennium: A Rationalist's Guide to a Precisely Arbitrary Countdown.* New York: Harmony.

Haines, A. and A. J. McMichael, eds. 1999. *Climate Change and Human Health* (Proceedings of a discussion meeting). London: Royal Society.

Hamblyn, Richard. 2001. *The Invention of Clouds*. New York: Farrar, Straus and Giroux.

Herschy, R. W. and R. W. Fairbridge, eds. 1998. *Encyclopedia of Hydrology and Water Resources*. Dordrecht: Kluwer Academic.

Hertz, J. H. 1967. *The Pentateuch and Haftorahs* (2d ed.). London: Soncino Press.

Hillairet, Jacques. 1963. *Dictionnaire Historique des Rues de Paris* (3d ed.). Paris: Editions de Minuit.

Hillel, Daniel. 1994. *Rivers of Eden: The Struggle for Water and the Quest for Peace in the Middle East*. New York: Oxford University Press.

Hoffman, P. F. and D. P. Schrag. 2000. Snowball Earth. *Scientific American* 282.1: 50–57.

Holloway, M. 2000. The killing lakes. *Scientific American* 283.1: 80–87.

Houze, Jr., Robert A. 1993. *Cloud Dynamics*. San Diego: Academic Press.

Howlett, Jr., Clifford T. 1997. *Chlorine and Chlorine Compounds in the Paper Industry*. Ann Arbor, Mich.: Ann Arbor Press.

Hoyle, Sir Fred, Geoffrey Burbidge, and J. V. Narlikar. 2000. *A Different Approach to Cosmology*. Cambridge: Cambridge University Press.

Humboldt, Alexander von. 1846. *Cosmos—essai d'une description physique du monde*. Translated by H. Faye (supervised by the author). Paris: Gide et Cie.

Imbrie, J. and K. P. Imbrie. 1979. *Ice Ages: Solving the Mystery*. Short Hills, N.J.: Enslow.

IPCC. 2001. Intergovernmental Panel on Climate Change, Third Assessment Report, Climate Change 2001. WG I: The Scientific Basis, Summary for Policy Makers; WG II: Impacts, Adaptation, and Vulnerability, Technical Summary.

Issar, Arie S. 1990. *Water Shall Flow from the Rock: Hydrogeology and Climate in the Lands of the Bible*. New York: Springer.

Jacobsen, T. 1958. Salt and silt in ancient Mesopotamian agriculture. *Science* 128: 1251–59.

Jacquet, Joseph., ed. 1997. L'eau et la vie des hommes au XXIème siècle. Paris: *Cahiers du MURS*, nos. 32–33.

Jaworowski, Zbigniew. 1999. Radiation risk and ethics. *Physics Today* 52.9: 24–29.

Jonas, Hans. 1984. *The Imperative of Responsibility: In Search of an Ethics for the Technological Age*. Translated by H. Jonas and D. Herr. Chicago: University of Chicago Press.

Joussaume, Sylvie. 1993. *Climat d'Hier à Demain*. Paris: CNRS-éditions.

Juteau, Thierry. 1993. *La Naissance des Océans*. Paris: Payot.

Kandel, Robert. 1992. *Our Changing Climate*. New York: McGraw-Hill.

——. 1998. *L'Incertitude des Climats*. Paris: Hachette-Pluriel.

Koeppel, Gerard T. 2000. *Water for Gotham: A History*. Princeton: Princeton University Press.

Kutzbach, Gisela. 1979. *The Thermal Theory of Cyclones: A History of Meteorological Thought in the Nineteenth Century*. Boston: American Meteorological Society.

Labeyrie, Jacques. 1993. *L'Homme et le Climat*. Paris: Denoël.

Larson, Erik. 1999. *Isaac's Storm: A Man, a Time, and the Deadliest Hurricane in History*. New York: Crown.

Lascar, J. and P. Robutel. 1993. The chaotic obliquity of the planets. *Nature* 361: 608–12.

Laskar, J., F. Joutel, and P. Robutel. 1993. Stabilization of the Earth's obliquity by the Moon. *Nature* 361: 615–17.

Laubier, Lucien. 1986. *Des Oasis au Fond des Mers*. Monaco: Sciences et Découvertes—Le Rocher.

——. 1992. *Vingt Mille Vies sous la Mer*. Paris: Ed. Odile Jacob.

Le Dren, J., ed. 1991. *Le Partage de l'Eau*. Revue *POUR*, nos. 127–128, Paris: L'Harmattan.

Le Pichon, Xavier. 1986. *Kaïko: Voyage aux Extrémités de la Mer*. Paris: Ed. Odile Jacob.

Levitus, S., J. I. Antonov, T. P. Boyer, and C. Stephens. 2000. Warming of the world ocean. *Science* 287: 2225–29.

Likens, Gene E. 1999. The science of nature, the nature of science: Long-term ecological studies at Hubbard Brook. *Proceedings of the American Philosophical Society* 143: 558–71.

Lorius, Claude. 1991. *Les Glaces de l'Antarctique, une Mémoire, des Passions*. Paris: Ed. Odile Jacob.

Lovejoy, Arthur O. 1936. *The Great Chain of Being: A Study of the History of an Idea*. Cambridge: Harvard University Press.

Lutz, Wolfgang, W. Sanderson, and S. Scherbov. 2001. The end of world population growth. *Nature* 412: 543–45.

Malin, M. C. and M. H. Carr. 1999. Groundwater formation of Martian valleys. *Nature* 397: 589–91.

Malin, M. C. and K. S. Edgett. 2000. Evidence for recent groundwater seepage and surface runoff on Mars. *Science* 288: 2330–35.

Margat, J. 1992. Gérer l'eau: Savoir et pouvoir. *La Science au Présent* 2: 298–302. Paris: Encyclopaedia Universalis.

Margat, J. and J.-R. Tiercelin, coords. 1998. *L'Eau en Questions: Enjeu du XXIe Siècle*. Paris: Romillat.

Marsily, Ghislain de. 1986. *Quantitative Hydrogeology*. San Diego: Academic Press.

Marsily, Ghislain de. 1995. *L'Eau*. Paris: Flammarion/Dominos.

Mason, B. J. 1971. *Physics of Clouds*. Oxford University Press.

——. 1976. Towards the understanding and prediction of climatic variations. *Quarterly Journal of the Royal Meteorological Society* 102: 473–98.

Matricon, Jean. 2000. *Vive l'Eau*. Paris: Gallimard.

McEvedy, Colin and Richard Jones. 1978. *Atlas of World Population History*. Hammondsworth and New York: Penguin.

McKay, C. P. 1999. Bringing life to Mars. *Scientific American Presents* 10.1.

McKibben, Bill. 1990. *The End of Nature*. Garden City, N.Y.: Doubleday Anchor.

Medvedev, Zhores (Jaurès) A. 1980. *Nuclear Disaster in the Urals* (1977). New York: Vintage.

Melosh, H. J., ed. 1997. *Origins of Planets and Life*. Palo Alto: Annual Reviews.

Meyer-Abich, Adolf. 1969. *Alexander von Humboldt*. Bonn: Inter Nationes.

Milankovitch, Milutin. 1920. *Théorie Mathématique des Phénomènes Thermiques Produits par la Radiation Solaire*. Paris: Gauthier-Villars.

Milliman, J. D. and R. H. Meade. 1983. World-wide delivery of river sediment to the oceans. *Journal of Geology* 91: 1–21.

Minnaert, M. 1954. *The Nature of Light and Color in the Open Air*. New York: Dover.

Minster, Jean-François. 1994. *Les Océans*. Paris: Flammarion/Dominos.

Miskovski, Jean-Claude. 1988. *Les Sédiments, Témoins du Passé*. Monaco: Le Rocher.

Moore, Thomas Gale. 2000. *In Sickness or in Health: The Kyoto Protocol Versus Global Warming*. Stanford: Hoover Institution Press.

Munitz, Milton K., ed. 1957. *Theories of the Universe: From Bablylonian Myth to Modern Science*. New York: Free Press.

Newton, Isaac. 1966. *Principia: Mathematical Principles of Natural Philosophy and System of the World* (1687). Translated by A. Motte, revised by F. Cajori, 1934. Rpt., Berkeley: University of California Press.

Nials, F. L., E. E. Deeds, M. E. Moseley, S. E. Pozorski, T. G. Pozorski, and R. Feldman. 1978. El Niño: The catastrophic flooding of coastal Peru. *Field Museum Bulletin* (Chicago) (July-August 1978): 4–14; Part 2 (September 1978): 4–10.

Nickson R., J. McArthur, W. Burgess, K. M. Ahmed, P. Ravenscroft, and M. Rahman. 1998. Arsenic poisoning of Bangladesh groundwater. *Nature* 395: 338.

Nitecki, Matthew. H, ed. 1981. *Biotic Crises in Ecological and Evolutionary Time*. New York: Academic Press.

NRC (National Research Council). 2001. *Under the Weather: Climate, Ecosystems, and Infectious Disease*. Commission on Climate, Ecosystems, Infectious Diseases, and Human Health, Board on Atmospheric Sciences and Climate, NRC. Washington, D.C.: National Academy Press.

O'Melia, Charles R. (Chair). 1999. *Watershed Management for Potable Water Supply: Assessing the New York City Strategy*. Water Science and Technology Board, National Research Council. Washington, D.C.: National Academy Press.

Orgel, L. E. 1973. *The Origins of Life: Molecules and Natural Selection*. New York: Wiley.

Pappalardo, R. T., J. W. Head, and R. Greeley. 1999. The hidden ocean of Europa. *Scientific American* 281.4: 34–43.

Pauling, Linus and Roger Hayward. 1964. *The Architecture of Molecules*. San Francisco: W. H. Freeman.

Pearce, F. 2001. Death in a glass of water. *The Independent* (England), January 19, 2001.

Pecker, Jean-Claude, coord. 1985. *L'Astronomie Flammarion*. Paris: Flammarion.

Petit, J. R. et al. 1999. Climate and atmospheric history of the past 420,000 years from the Vostok ice core, Antarctica. *Nature* 399: 429–36.

Philander, George. 1990. *El Niño, La Niña, and the Southern Oscillation*. San Diego: Academic Press.

Pinker, Steven. 1995. *The Language Instinct*. London: Penguin.

Plaut, Steven. 2000. *Water Policy in Israel*. Policy Study no. 47. Jerusalem and Washington, D.C.: Institute for Advanced Strategic and Political Studies.

Popper, Karl R. 1963. *Conjectures and Refutations: The Growth of Scientific Knowledge*. London: Routledge and Kegan Paul.

Postel, S. L., G. C. Daily, and P. R. Ehrlich. 1996. Human appropriation of renewable fresh water. *Science* 271: 785–88.

Press, Frank and Raymond Siever. 1986. *Earth* (3d ed.). San Francisco: W. H. Freeman.

Rasmussen, E. M. 1967. Atmospheric water vapor and water balance of North America. *Monthly Weather Review* 95: 403–26.

Rasool, I. and C. De Bergh. 1970. The runaway greenhouse effect and the accumulation of CO_2 in the Venus atmosphere. *Nature* 226: 1037–39.

Rescher, Nicholas. 1995. *Luck: The Brilliant Randomness of Everyday Life*. New York: Farrar Straus Giroux

Riser, Jean. 1995. *Erosion et Paysages Naturels*. Paris: Flammarion/Dominos.

Roan, Sharon L. 1990. *Ozone Crisis*. New York: Wiley.

Robinson, Kim Stanley. 1993. *The Mars Trilogy—Red Mars, Green Mars, Blue Mars*. New York: Bantam.

Rosenfeld, Arthur H., Tina M. Kaarsberg, and Joseph Romm. 2000. Technologies to reduce carbon dioxide emissions in the next decade. *Physics Today* 53.11: 29–34.

Rosenfeld, Daniel and I. M. Lensky. 1998. Satellite-based insights into precipitation formation processes in continental and maritime convective clouds. *Bulletin of the American Meteorological Society* 79: 2457–76.

Roux, Frank. 1991. *Les Orages—Météorologie des Grains, de la Grêle et des Eclairs*. Paris: Payot.

Ryan, William and Walter Pitman. 2000. *Noah's Flood: The New Scientific Discoveries About the Event That Changed History*. New York: Simon and Schuster–Touchstone.

Salem, Lionel. 1990. *Le Dictionnaire des Sciences*. Paris: Hachette.

Schneider, S. H. and L. E. Mesirow. 1976. *The Genesis Strategy: Climate and Global Survival*. New York: Plenum.

Scorer, Richard and Arjen Verkaik. 1989. *Spacious Skies*. Newton Abbot, London: David and Charles.

Shklovskii, I. S. and Carl Sagan. 1966. *Intelligent Life in the Universe*. New York: Holden-Day.

Simon, Julian L. and H. Kahn. 1981. *The Resourceful Earth*. New York: Basic Books.

Smith, Robert A. 1872. *Air and Rain: The Beginnings of a Chemical Climatology*. London: Longman and Green.

Stahle, David W., M. K. Cleaveland, D. B. Blanton, M. D. Therrell, and D. A. Gay. 1998. The lost colony and Jamestown droughts. *Science* 280: 546–67.

Stanley, Steven M. 1989. *Earth and Life Through Time* (2d ed.). San Francisco: W. H. Freeman.

Starr, Joyce Shira. 1995. *Covenant Over Middle Eastern Waters: Key to World Survival*. New York: Holt.

Stewart, R. E., K. K. Szeto, R. F. Reinking, S. A. Clough, and S. P. Ballard. 1998. Mid-latitude cyclonic cloud systems and their features affecting large scales and climate. *Reviews of Geophysics* 36: 245–74.

Tamir, Avraham. 1998. *A Soldier in Search of Peace: An Inside Look at Israel's Strategy*. London: Weidenfeld and Nicolson.

Tardy, Yves. 1986. *Le Cycle de l'Eau: Climats, Paléoclimats, et Géochimie Globale*. Paris: Masson.

Thornton, Joe. 2000. *Pandora's Poison: Chlorine, Health, and a New Environmental Strategy*. Cambridge: MIT Press.

Thoreau, Henry David. 1897. *Walden; Or, Life in the Woods* (1854). Cambridge, Mass.: Riverside Press.

Timberlake, Lloyd. 1985. *Africa in Crisis: The Causes, the Cures of Environmental Bankruptcy*. Washington, D.C.: Earthscan.

Tubiana, L. and Benjamin Dessus, eds. 1998. Le climat, risque majeur et enjeu politique: De la conférence de Kyoto à celle de Buenos Aires. *Courrier de la Planète/Cahiers de Global Chance* (Paris) (March-April).

UNESCO (United Nations Educational, Scientific, and Cultural Organization). 1978. *World Water Balance and Water Resources of the Earth*. Paris: UNESCO.

Velikovsky, Immanuel. 1950. *Worlds in Collision*. New York: Macmillan.

Villiers, Marq de. 2000. *Water: The Fate of Our Most Precious Resource*. New York: Houghton Mifflin.

Vitousek, P. M., H. A. Mooney, J. Lubchenco, and J. M. Melillo. 1997. Human domination of Earth's ecosystems. *Science* 277: 494–99.

Vlachos, Evan, ed. 1990. *Water, Peace, and Conflict Management*. Special issue, *Water International* 15.4: 185–239.

Vogt, William. 1948. *Road to Survival*. New York: William Sloane.

Vörösmarty, Charles J., P. Green, J. Salisbury, and R. B. Lammers. 2000. Global water resources: Vulnerability from climate change and population growth. *Science* 289: 284–88.

Wallace, J. M. and P. V. Hobbs. 1977. *Atmospheric Science: An Introductory Survey*. San Diego: Academic Press.

Ward, W. R. 1973. Large-scale variations in the obliquity of Mars. *Science* 181: 260–62.

Watson, I. and A. D. Barnett. 1995. *Hydrology, an Environmental Approach*. Boca Raton: CRC Lewis.

Watson, James D. 1981. *The Double Helix: A Personal Account of the Discovery of the Structure of DNA* (1968). New York: Atheneum; London: Weidenfeld and Nicholson.

Wei, Huai Fu, E. Ledoux, and G. de Marsily. 1990. Regional modeling of groundwater flow and salt and environmental tracer transport in deep aquifers in the Paris basin. *Journal of Hydrology* 120: 341–58.

Weinberg, Steven. 1993. *The First Three Minutes: A Modern View of the Origin of the Universe*. New York: Basic Books.

Weizsäcker, Ernst von, Amory B. Lovins, and L. Hunter Lovins. 1997. *Factor Four— Doubling Wealth, Halving Resource Use: The New Report to the Club of Rome*. London: Earthscan.

Wells, Herbert George. 1978. "The Man Who Could Work Miracles." In Wells, *Selected Short Stories* (1927), 299–315. London: Penguin.

Wilson, Edward O. 1998. *Consilience: Unity of Knowledge*. New York: Knopf.

Woolf, Aaron T. 1995. *Hydropolitics Along the Jordan River*. New York, Paris, Tokyo: United Nations University Press.

Zachar, Dusan. 1982. *Soil Erosion*. New York: Elsevier.

Zorpette, G. 1999. A probing prelude—Global Surveyor bolsters the theory that water persisted on Mars. *Scientific American* 280.4: 15.

New York (city, state): 77, 172; climate
and rainfall, 89, 150, 157–58;
epidemics: 200, 213; water
supply, 203–206, 278nn9&12,
279nn16&17, 280n25, 288n1;
water pollution: 206–207;
wetlands, 212
Niger River, 157, 160, 191
Nile River, 100, 155, 160, 177–79,
195–96, 209, 231, 233–34,
270n17, 283n12
Niña, La; Niño, El. See El Niño
nitrates, 170, 203, 242
nitrogen: 17; in the atmosphere: 3, 5,
7, 24, 27, 42, 52, 90–93, 98, 242;
C-14 production, 128; organic
molecules, 40, 43; nitrogen
oxides, 153
nomenclature: chemical elements, 3;
clouds, 135
North America, 120, 122; ice age, 81,
172; relief and rivers, 144, 155,
160, 179–81; water availability
and use, 208, 222–30
North Atlantic Oscillation. See NAO
Norwegian Sea, 122, 131–32, 139, 261
nuclear physics, 11–17, 25
nuclear power, 208, 216–19, 231, 255,
281nn29&30

oases, 165, 256, 276n3
obliquity of the ecliptic, 70, 74–75, 79,
270nn14&15
oceans: 25, 28–29, 48, 85, 87–88, 97,
102, 118, 125–34, 140–41; and
CO_2, 50, 58; currents, 104–108,
133; deep water, 7, 112–15,
125–31; ice, 77, 259, 266n10; and
life, 62; on other planets, 29, 75,
265–66nn3&4; salinity, 60;
tropical cyclones, 150; warming,
5, 273nn8.5&7; in the water cycle,
7, 27–28, 97–101, 159, 249. See
also conveyor-belt circulation;

currents; sea-floor spreading;
sea level; tides; tsunami; see also
by name
oceanic trench, 29–33, 36–38. See also
sea-floor spreading
Ogallala aquifer, 125, 157, 227
Ohio River, 179
oil, xiii, 51, 54, 128, 168, 232–33,
242–45, 253, 255, 273n8, 287n2
orbit: Earth, 20, 73–75, 79; other
planets, comets, 20–24, 64–65,
67–75, 79, 95, 97; artificial
satellites, 68, 184, 263n1(pref),
265n2.3, 274n11
organochlorines, 201, 279n18
origin: galaxies and stars, 14–16,
20–22; life 18, 39–41, 45–50; solar
system and Earth, 23–26, 29
oscillation. See Southern
Oscillation; NAO
oued, 170
outgassing of atmosphere and water,
25–28, 261
oxygen: 3–7, 11, 43–44, 49–55; anoxic
(oxygen-free) environments,
45–47, 52, 55, 243; atmospheric,
24, 27, 45–47, 52–53, 58, 92–93;
isotopes, 76–78, 99, 124; in
organic molecules, 40;
photosynthesis, 7, 50, 52–54, 99,
161, 224; production in stars,
14–18
ozone: depletion, 129, 239; formation,
53; as pollutant, 153; as ultraviolet radiation shield, 92–93;
in water purification, 203

Pacific Ocean, 100, 105–108, 111–28,
157, 190, 241. See also El Niño;
sea-floor spreading; tsunami;
typhoons
paleoclimatology, 73, 76–82, 124,
270n18, 277n8. See also isotopes
Paleozoic, 46, 57